POWER TOOL SAFETY AND OPERATION

INTRODUCTION

This manual, **POWER TOOL SAFETY AND OPERATION**, has been prepared for instructors and students in agricultural education, industrial education, trades, industry, and other agencies for safety instructional programs in the area of power tool safety and operation. In addition, this manual would be of real value to the home owner aiding in the safe as well as proper operation of power tools and equipment commonly found in the home shop.

The manual is designed to give the student a basic understanding of the proper operation of power tools. Thirty power tools are covered in the areas of woodworking, metalworking, and metals and welding. For each power tool, there is a section on part identification, safe operational procedures, general safety practices, and completion questions. After the student has read the written material and completed the part identification and completion question section, the instructor should determine the proper adjustment and operational features of the power tool being studied. Following this, the student should successfully complete the safety exam and then be allowed to operate the power tool with close supervision from the instructor. The next step in this educational process would be for the student to complete an activity of his/her own choosing using the power tool or tools that have been studied. NOTE: As manufacturers' designs differ for the same power tool, make certain the operator's manual is available and used for making adjustments and operation of the particular power tool.

A Teacher Answer Key for part identification and completion questions is available. Also, an Instructor's Packet designed to be used in conjunction with this manual is available from Hobar Publications. It contains suggested teacher and student activities, a transparency master set of part identification, and the safety exams. This packet is very essential in the teaching of this unit.

Almost everyone involved in school laboratory instructional programs will come in contact at some time with power tools and equipment. The safe as well as proper operation of these tools are essential for quality and efficient workmanship. Knowledge on the safe and proper operation of power tools will make the work easier, more enjoyable, and increase operator satisfaction and job performance.

THE AUTHORS

THOMAS A. HOERNER
Professor Emeritus
Dept. of Agriculture & Biosystems Engineering
and Agriculture Education & Studies
Iowa State University
Ames, Iowa

MERVIN D. BETTIS
Professor Emeritus
Department of Agriculture
Northwest Missouri State University
Maryville, Missouri

HOBAR PUBLICATIONS
A DIVISION OF FINNEY COMPANY
8075 215th Street West
Lakeville, Minnesota 55024
Phone: (952) 469-6699 or (800) 846-7027
www.finney-hobar.com

ISBN 13: 978-0-913163-30-6
Copyright © 1998 by Hobar Publications

All rights reserved. No part of this book covered by the copyrights hereon may be reproduced or copied in any form or by any means—graphic, electronic, or mechanical, including photocopying, taping, or information storage and retrieval systems—without written permission of the publishers.

First Printing	1973
Second Printing	1975
Third Printing	1977
Fourth Printing	1980
Fifth Printing	1984
First Revision	1987
Seventh Printing	1990
Eighth Printing	1994
Second Revision	1998
Tenth Printing	2001
Eleventh Printing	2005
Twelfth Printing	2008

TABLE OF CONTENTS

TITLE **PAGE**

Title	Page
Bench Saw	1
Jig Saw	5
Band Saw	9
Sabre Saw	13
Radial Arm Saw	17
Portable Circular Saw	21
Reciprocating Saw	25
Power Miter Saw	29
Finishing Sander	33
Belt Sander	37
Disc Sander	41
Wood Lathe	45
Planer-Surfacer	49
Portable Power Planer	53
Shaper	57
Router	61
Jointer	65
Drill Press	69
Portable Drill	73
Nailer and Stapler	77
Grinder	81
Belt Grinder	85
Portable Grinder	89
Metal Cutting Band Saw	93
Portable Metal Cutting Band Saw	97
Metal Lathe	101
Air Impact Wrench	105
Gas Forge	109
Arc Welder	113
Oxyacetylene Welder	117

BENCH SAW

A. **PART IDENTIFICATION:**

Identify the numbered parts of the bench saw illustrated below.

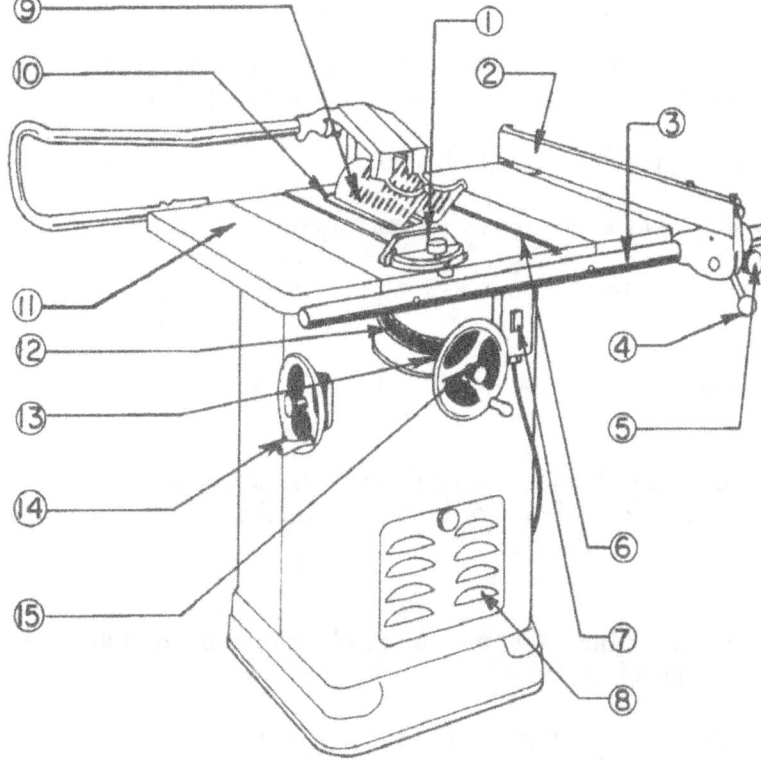

1. _____
2. _____
3. _____
4. _____
5. _____
6. _____
7. _____
8. _____
9. _____
10. _____
11. _____
12. _____
13. _____
14. _____
15. _____

B. **SAFE OPERATIONAL PROCEDURES:**

1. Crosscutting:

    a. Use a crosscut or combination blade. Always use sharp blades. Make sure blade is always cutting down toward front or operator's side of table.

    b. Adjust the depth of cut so that the teeth clear the thickness of material by the depth of the teeth.

    c. Be sure saw guard and splitter are in place. Check anti-kickback device to make sure it is working.

    d. Always use the miter gage when crosscutting. A wood facing is recommended for the miter gage.

    e. Never use the ripping fence as a guide when crosscutting short pieces. Use a stop on the miter gage or stop blocks clamped to the ripping fence or the table top.

      f.   Place board against miter gage and saw the board, saving the cutting line.

      g.   Do not force work through the saw.

2. Ripping:

    a.   Use a ripping or combination blade.

    b.   Use ripping fence as a guide. <u>Never</u> saw freehand.

    c.   Double check cutting width by measuring from fence to outer teeth. Also check front and back of blade so fence will not bind material to blade.

    d.   When ripping narrow pieces, use a push stick.

    e.   Be sure guards are in place for all sawing operations.

    f.   Use a helper or a roller stand to support long pieces of material while sawing.

    g.   Adjust blade to proper height, 1/4" to 1/2" above the material being sawed.

    h.   Do not force material into blade. If blade overheats, stop saw immediately and check for dullness or binding of blade.

3. Dadoing or plowing:

    a.   Use a dado blade if available. A combination blade may be used and the fence varied with each piece.

    b.   Raise the saw to the desired depth of cut. Make pass on waste piece before cutting into actual work piece.

    c.   This operation may require the guard to be removed if the material is to be dadoed on edge.

4. Bevel cuts (ripping):

    a.   Use a combination or ripping blade.

    b.   Adjust saw by tilting the arbor to the angle desired.

    c.   Adjust blade to proper height, 1/4" to 1/2" above work.

    d.   Adjust ripping fence to desired width of cut.

5. Bevel cuts (crosscutting):

    a.   Use a combination or crosscutting blade.

    b.   Use miter gage (not the ripping fence) and follow the steps as listed in Item 4.

C. **GENERAL SAFETY PRACTICES:**

1. Wear eye protection, hearing protection, and proper clothing when operating this machine.

2. Obtain permission from instructor to operate bench saw.

3. Use only sharp blades of the proper type for the job.

4. Be sure blade is correctly installed in the saw. Make certain power is off and properly disconnected before removing blade insert. To loosen arbor nut, turn wrench toward normal direction of travel holding blade with waste piece of wood. Do not overtighten arbor nut when replacing blade.

5. Do not stand in line with the blade while sawing or allow fingers or hands to be in line of cut.

6. Be sure that all adjustments are tight and the table part of the saw is free of tools, chips, small pieces of wood, or other materials.

7. Do not talk to anyone while using the saw. The operator should be the only person inside the safety zone.

8. Be sure the floor is clean and free from scraps and rubbish. Do not work on wet or slippery floors. Non-skid materials are recommended.

9. Saw only material that has a straight edge. To avoid chipping of material such as plywood or masonite, saw with face grain or good surface down.

10. Study the adjustments and make sure you understand them before starting to work.

11. Use the saw guard where possible.

12. Use a "push" stick for ripping narrow pieces.

13. Hold material against the ripping fence when ripping, and the miter gage when crosscutting. Never saw freehand.

14. Never use the ripping fence for a gage when crosscutting short pieces.

15. Do not place the hands over or in front of the blade. Never reach over the blade.

16. Turn off the saw before removing short pieces from near the blade.

17. Be sure the power is "locked" off before adjusting or working on the saw.

D. **COMPLETION QUESTIONS:**

1. A _____ or _____ _____ is used to support long pieces of material while sawing.

2. The _____ _____ is used for a guide when ripping.

3. The _____ _____ is used as a guide when crosscutting.

4. A _____ or _____ blade may be used for crosscutting.

5. A _____ _____ is used when ripping narrow pieces.

6. When making a bevel cut, the saw is adjusted by tilting the _____.

7. Saw only material that has a _____ edge.

8. The saw should be _____ before removing short pieces from the blade.

9. A _____ or _____ blade can be used when plowing or dadoing.

10. The blade should extend above the material the depth of one _____.

JIG SAW

A. PART IDENTIFICATION:

Identify the circled parts on the jig saw illustrated below.

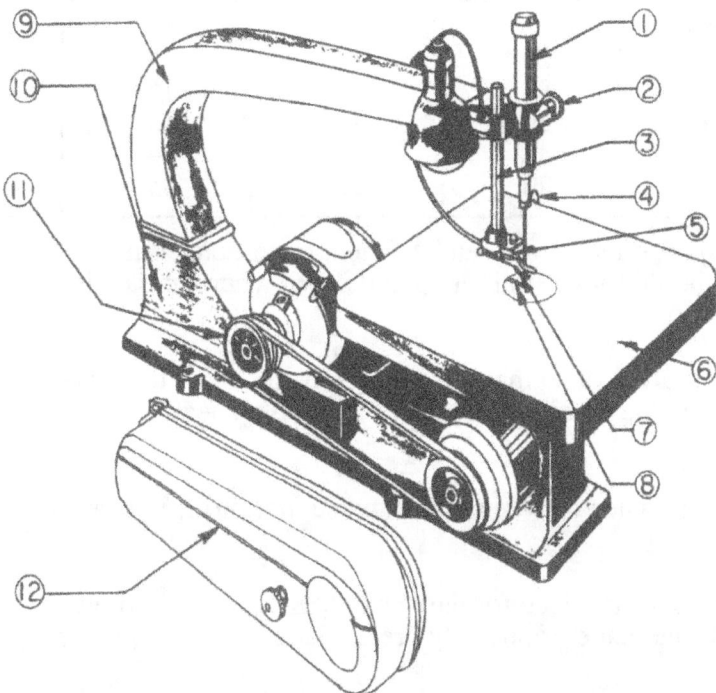

1. _____
2. _____
3. _____
4. _____
5. _____
6. _____
7. _____
8. _____
9. _____
10. _____
11. _____
12. _____

B. SAFE OPERATIONAL PROCEDURES:

1. Select blade according to type of job. There are two general types of blades.

    a. Jeweler's blade - This blade is held in both the upper and lower chuck and is used for fine work.

    b. Sabre blade - This blade is held only by the lower chuck and is for heavier cutting.

2. Regarding teeth per inch, blades vary from 7 teeth for rough cutting to 32 teeth recommended for cutting metal or other hard materials. Blades having 15 teeth per inch are recommended for general purpose cutting.

3. Insert blade with the teeth pointing down so that cutting is on the downward stroke, thus holding the work against the table.

4. After inserting blade, tighten chuck properly and then adjust blade tension with tension sleeve, keeping the blade taut.

5. Determine proper speed according to job. The cutting strokes per minute can be regulated by changing the belt on the cone pulleys or by adjusting variable speed drive if saw is equipped with a multi-speed drive. Be sure guard is replaced before operating saw. Note blade and speed selection chart below:

Material to be cut	Speed	Thickness to be cut			
		up to 1/16"	1/16"-1/4"	1/4"-1/2"	1/2"-2"
		Number of teeth per inch on blade			
Hardwood	1000-1750	20	16	15	10
Softwood	1750	20	18	15	8
Plywood	1300-1750	20	18	15	10
Plastics	650- 900	20	18	15	12
Steel	650	32	20	—	—
Aluminum	650- 900	20	20	15	—

6. Blade alignment is very important for safe and efficient operation. When viewed from the side, the blade should move straight up and down when the saw is running.

7. Adjust the hold-down pressure foot so that the spring tension holds the work tight to the table. Turn saw one revolution by hand to double check all adjustments.

8. Note size of throat opening (space between blade and support arm) in respect to size of piece to be cut.

9. Before beginning cut, check work piece for nails, paint, grit, or other foreign material which could dull the blade or possibly break blade.

10. With table clear of all chips, small pieces, or other materials, start the machine and feed the work into the blade forward and evenly with a slight downward pressure. Saw cuts should be on the waste side of the line. The line should be barely visible on work after cutting.

11. The jig saw can be used for inside cuts.

    a. Drill small holes in the waste stock at points of abrupt changes of direction on curves.

    b. Insert the blade through the drilled hole and secure in chuck as discussed earlier.

    c. Proceed with cut as with an outside cut.

    d. When finished with cut, stop machine, raise upper guide, and remove blade from upper chuck to free the work piece.

12. Do not force the work into blade or attempt to turn too sharply. If blade should break, turn the saw off and allow the blade to come to a complete stop before removing work or replacing blade.

13. Bevel cuts can be made if the saw is equipped with a tilt table.

14. If saw is equipped with a light and dust blower, be sure both are working properly before operating the saw.

C. GENERAL SAFETY PRACTICES:

1. Wear eye protection and proper clothing when operating this saw.
2. Do not operate saw without instructor's permission.
3. Use only sharp blades.
4. Select correct speed and type of blade for work to be done.
5. Keep work area clear and uncluttered.
6. All guards should be in place at all times when operating the saw.
7. Make all adjustments before starting machine.
8. Pressure should be held on work by the hold-down foot.
9. Keep hands and fingers out of line of the saw.
10. Do not force material into the saw or attempt to turn too short a radius.

D. COMPLETION QUESTIONS:

1. The saw should be operated at approximately _____RPM with a blade having _____ teeth per inch when cutting a piece of 3/4" pine lumber.

2. If saw cuts are rough and ragged, the speed should be _____.

3. When using the _____ type blade, both chucks are used.

4. The work is held snug to the table by the _____ - _____ _____.

5. For general purpose sawing, a blade having _____ teeth per inch is recommended.

6. The blade should be inserted so that sawing is completed on the _____ stroke.

7. When using the sabre saw blade, only the _____ chuck is used.

8. Proper blade tension is accomplished by adjusting the _____ _____.

9. To increase speed if the saw is designed with a 4-step motor and drive pulley, the belt should be moved to a _____ position on the motor and a _____ position on the drive cone pulley.

10. Saw cuts should be made on the _____ side of the line.

NOTES

BAND SAW

A. PART IDENTIFICATION:

Identify the parts on the band saw illustrated below:

1. _____
2. _____
3. _____
4. _____
5. _____
6. _____
7. _____
8. _____
9. _____
10. _____
11. _____
12. _____
13. _____
14. _____
15. _____
16. _____

B. SAFE OPERATIONAL PROCEDURES:

1. Adjust the upper guard and guide about 1/8" to 1/4" above material to be cut.

2. Select proper blade width. No cutting radius should be too small for the blade. General rules regarding minimum radius cuts by blade width:

Blade width	Min. Radius
3/4"	1-3/4"
1/2"	1-1/4"
3/8"	1"
1/4"	3/4"
3/16"	1/2"
1/8"	1/4"

3. Keep blades sharp and properly set. If blade leads to one side, it may be dull, unevenly set, or a guide may be improperly set. Blade guides should be snug but not tight against the blade (paper thickness on each side).

4. Teeth on blade should point down or in direction of travel of blade.

5. Check tension of the blade frequently. Make adjustments as necessary. Check adjustment of blade by hand-turning the band wheels.

6. Adjust band saw blade so that it will run on the center of the wheels and straight through the guides.

7. When not cutting, the blade should run 1/32 of an inch ahead of the blade support wheel. If the rear edge of blade runs on the wheel, it will become case hardened and possibly break.

8. Be sure work is free of nails, paint, and obstructions.

9. Mark the material clearly so that lines can be seen at a reasonable distance.

10. Plan cuts to avoid backout from curves when possible.

11. Plan cuts and lay out. Make "release" cuts before cutting long curves.

12. If saw has variable speed, operate blade speed at approximately 3000 feet per minute for wood.

13. Do not start cut until the saw is running at full speed.

14. Feed the stock into the saw slowly and only as fast as the saw will take it readily.

15. If freehand sawing, use one hand to guide the work and the other to push the work into the saw.

16. Do not force or twist the wood into the blade so as to produce excess stress on the blade.

17. If the stock binds or pinches the blade, do not attempt to back out until the power has been shut off and the machine stops.

18. Always turn off the machine before the blade is backed out of a cut.

19. Use the fence guide bar or miter gage when sawing straight lines.

20. The table should be tilted for making bevel cuts.

21. Cylindrical stock should be mounted in a holding device to keep it from spinning and crowding the blade.

22. Remove all scraps and chips from the table with a brush only when the power is off and the blade is stopped.

C. **GENERAL SAFETY PRACTICES:**

1. Wear safety glasses at all times when operating a band saw.

2. Do not operate saw without instructor's permission.

3. Avoid wearing loose clothing.

4. Keep floor and surrounding area free of scrap that might cause tripping.

5. Be sure saw is properly grounded.

6. Keep all guards in place at all times.

7. Get someone to assist in operations which are not safely handled alone.

8. Make all adjustments with the power off and blade stopped.

9. Keep hands a safe distance from moving parts, never closer than 2 inches from the blade.

10. Give undivided attention to the job. The operator should be the only one inside the safety zone area.

11. Use a push block when sawing small stock.

12. Never reach around a moving blade.

13. When making a cut, do not place hands in line with the cutting edge.

14. Never attempt to remove small pieces of wood from near the blade while the saw is running.

15. Never leave a running saw unattended.

16. When finished cutting, shut off the switch and disconnect machine from power source. Do not leave the safety zone until the blade comes to a complete stop.

D. COMPLETION QUESTIONS:

1. Use only a sharp blade with proper _____.

2. When the blade is properly installed, the teeth should point _____.

3. The blade should be adjusted so it runs _____ of an inch from the blade support wheel.

4. Keep hands at least _____ inches from the blade.

5. Use a _____ _____ when sawing small stock.

6. Adjust the upper guide about _____ inches above the material being cut.

7. Properly adjust the band saw blade so it will run on the _____ of the wheels.

8. Move the upper guide only while the saw is _____.

9. The smallest radius that can be safely cut with a 1/2" wide blade is _____ inches.

10. A dull or improperly set blade will cause the blade to _____ to one side on the work piece.

NOTES

SABRE SAW

A. PART IDENTIFICATION:

Identify the circled parts on the sabre saw illustrated below.

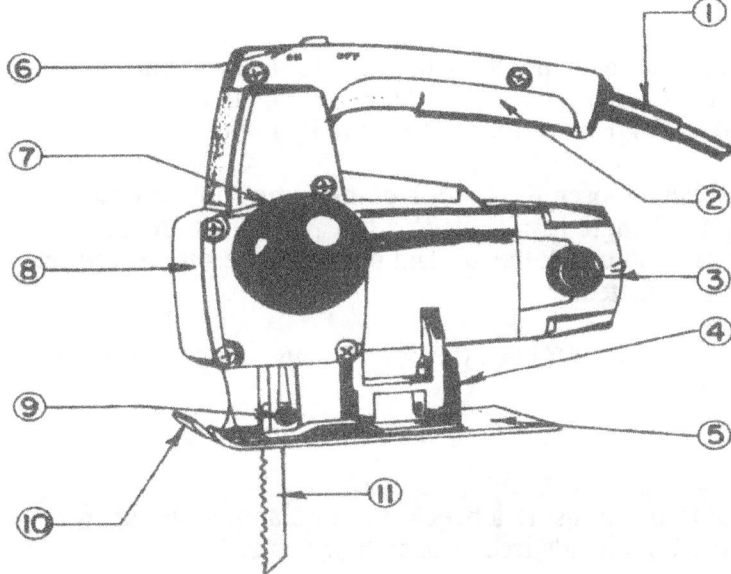

1. _____
2. _____
3. _____
4. _____
5. _____
6. _____
7. _____
8. _____
9. _____
10. _____
11. _____
12. _____

B. SAFE OPERATIONAL PROCEDURES:

1. Replacing the sabre saw blade:

    a. Many blades are available for use in the sabre saw. Be sure the blade you use is the type for the work. Blade types are the wood-cutting, metal-cutting, and knife blades. There are also many blades available within each blade type. Follow manufacturer's blade recommendations for each job. Note: Two (2) teeth or more should be in contact with the cutting surface at a time.

    b. Be sure the blade has the correct shank. One blade shank will not fit all brands of saws or even all models of saws within one brand.

    c. To remove blade, loosen blade screw and pull blade from slot. Turn screw back in several turns to prevent loss.

    d. To place a blade in the saw, loosen blade screw and insert blade shank until seated or hole in shank is aligned with blade screw. Teeth must be facing forward and pointing upward, then tighten screw firmly.

    e. Make no adjustments on saw while it is plugged into power source.

2. Adjustment of the sabre saw:

   a. Some sabre saws have a base which may be raised and lowered to vary the amount of exposed blade. The blade must project through the work when the blade is at the highest point of its stroke. Disconnect saw from power source before making any adjustments.

   b. To adjust the amount of exposed blade, loosen the vertical adjustment locking screw and raise or lower base to the required height; tighten locking screw firmly.

   c. Many sabre saws have bases which may be tilted for beveled cuts. These saws characteristically have an arc divided in 5, 10, or 15 degree increments located to the front or rear of the aluminum housing.

      (1) For beveled cuts, loosen the bevel locking screw, tilt the base until the desired angle is indicated on the arc, tighten the bevel locking screw, and check the arc and pointer to be sure the correct angle is still indicated.

      (2) For perpendicular or 90° cuts, the pointer must indicate 0 degrees of angle on the arc.

3. Operating the sabre saw:

   a. Secure the material to be cut using a bench vise or clamp to workbench or sawhorses leaving both hands free to operate the saw.

   b. The line of cut must be free from obstructions above and below the work.

   c. Be sure proper blade for the job is locked securely in blade holder, teeth forward and pointed up.

   d. Be certain switch is in "off" position before connecting to power source.

   e. The saw handle is grasped in the right hand and the guide knob in the left.

   f. If the saw has different operating speeds, determine the proper speed before beginning the saw cut.

   g. To start an outside cut, place the toe of the saw base on the edge of the material, start the motor, and move the blade into the work. Push forward and downward with the right hand and guide the saw with the left hand.

   h. Use constant, firm pressure on the saw to maintain a uniform forward movement. Do not overload the saw because it will result in an overheated motor, overheated blade, or blade breakage.

   i. Do not attempt to cut curves so sharp that the blade will be twisted. Use narrow-bodied blades for curves and wide-bodied blades for straight cuts.

   j. When making pocket or internal cuts, (1) drill a starting opening in the waste stock to begin the cut or (2) make a plunge cut by resting the toe of the base firmly on the work with the blade raised off the work, turn on the motor, and then lower the blade slowly into the work. Note: The plunge cut method is not recommended for materials over 1/2" in thickness.

k. For long cuts, the switch may be locked in the "on" position by turning the saw on and depressing the switch lock with the right thumb. To release the lock, pull up on the off-on switch. (This applies only to saws with a switch lock.)

l. To prevent binding or breakage of the blade, support the cut-off material until the cut has been completed.

m. When the cut has been completed, turn off the motor and set the saw down after the motor has stopped completely.

n. If the saw is to be removed from the cut prior to reaching the edge of the work, turn off the motor and wait until it has completely stopped before removing saw from cut.

o. When finished using the sabre saw, disconnect saw from power source and return saw to the storage chest. Depending on storage method, it may be recommended to remove the saw blade before storing the saw.

C. GENERAL SAFETY PRACTICES:

1. Never operate sabre saw without safety glasses or goggles.

2. Do not operate saw without instructor's permission.

3. Clamp work securely to prevent movement or excessive vibration.

4. Have plenty of shadow-free light on the work.

5. Check blade for proper size, sharpness, and tightness in blade holder. Do not use dull, bent, or cracked blades.

6. After making certain the switch is in the "off" position, connect power cord to power source. Saw must be properly grounded to prevent injury to user.

7. Maintain a well-balanced position on both feet. Do not shift position of feet while sawing.

8. Grip handle firmly with the right hand and control turning movements with left hand on the guide knob.

9. Place toe of base firmly on work before turning on motor.

10. Do not set saw down or remove blade from an unfinished cut until the motor is stopped.

11. When finishing a cut, support the cut-off section if it may bind the blade. Bound blades may break, throwing metal pieces some distance.

12. Always disconnect from power source when inspecting parts, making adjustments, and removing or installing blades.

D. **COMPLETION QUESTIONS:**

1. The vertical adjustment locking screw is used to adjust the amount of _____ exposed.

2. An angle of _____ degrees should be indicated on the arc for cuts perpendicular to the surface of the material.

3. The _____ hand should be used to grasp the _____ to guide the saw.

4. The types of blades are the _____, _____, and _____.

5. The amount of blade exposed below the base should be sufficient to project below the lower surface of the work at the _____ point of the saw blade stroke.

6. The _____ _____ holds the blade in place on the saw.

7. The _____ hand should be placed on the _____ and used to exert a constant forward and downward pressure on the saw.

8. When the blade is properly installed, the teeth must point _____ and _____.

9. When starting an outside cut, the _____ of the saw should be resting on the material before turning on the motor.

10. A type of internal cut called the _____ cut is started by resting the _____ of base firmly on the work with the blade raised above the work, and then turning on the motor and slowly lowering the blade into the work.

RADIAL ARM SAW

A. PART IDENTIFICATION:

Identify the circled parts on the radial arm saw illustrated below.

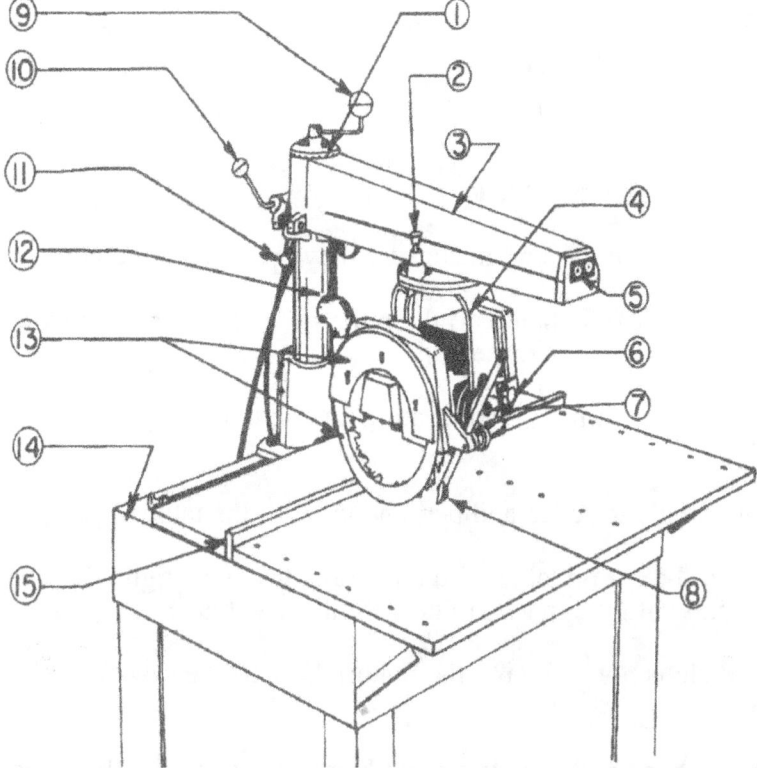

1. _____
2. _____
3. _____
4. _____
5. _____
6. _____
7. _____
8. _____
9. _____
10. _____
11. _____
12. _____
13. _____
14. _____
15. _____

B. SAFE OPERATIONAL PROCEDURES:

1. Changing the saw blade:

    a. Shut off main power switch or disconnect from wall plug.

    b. Select the proper type blade for job.

    c. Remove the saw guard, the arbor nut, and arbor collars. Remember the arbor has left-hand threads. Remove the blade.

    d. Before replacing new blade, check the inside collar. It should be replaced on the arbor with the recessed side out.

    e. Place the saw blade on the arbor so the teeth on the bottom of the blade or nearest the table point back or toward the elevating column. Generally, blades are marked "this side out."

    f. Place the outside collar on the arbor with the recessed face against the saw blade.

g. Tighten the arbor nut using the same wrenches used in removing.

h. Before replacing the guard, check the squareness of the saw blade and table by resting a framing square against the face of the blade and table top. If out of adjustment, consult the operator's manual for the saw.

i. Assemble the guard kickback and elbow.

j. Replace the guard and lock in place.

k. While the saw is still disconnected from electricity, rotate the saw blade by hand to see that it runs clear and free.

2. Crosscutting:

a. Select a crosscut or combination blade.

b. Have all guards in proper place and make sure they are free to operate.

c. Push the saw to the rear of the table. Tighten the rip lock to keep the saw from running forward when it is turned on.

d. Adjust height of blade by turning the elevating handle until the teeth just touch the table top.

e. Adjust saw at right angle to fence and perpendicular to the table.

f. Place the material to be cut on table with the straightest edge tight against the fence and with the mark in line with the saw blade.

g. Be sure the saw blade is not engaging the material. Start the saw and release the rip lock.

h. Hold wood with left hand and saw with right hand, standing slightly to the left of the line of the saw blade.

i. Pull saw toward operator feeding slow enough that the saw does not grab.

j. Return the saw to the rear of the table. Lock in place until ready to make another cut.

3. Ripping:

a. Select ripping or combination blade.

b. Turn saw parallel to fence by releasing the swivel lock and turning the saw yoke. Lock in position with rip lock at proper width of cut.

c. Adjust safety guard to approximately 1/8" above the board and the anti-kickback device at 60° when touching material to be ripped.

d. Adjust height of saw with elevating handle so that teeth just touch table.

e. Feed material from opposite end of anti-kickback device or into the bottom of the saw blade which is turning toward the operator.

f. Use a push stick when near blade to keep hands away from blade.

g. Get an assistant to help with long material or use a roller support.

4. Miter cuts:

   a. Select a crosscut or combination blade.

   b. Set the motor yoke and the lock in the same position as for crosscutting. Release the arm clamp and the miter latch.

   c. Swing the radial arm to the desired angle as indicated on the miter scale. Most saws have a notch or hole at 30°, 45°, 60°, etc. so that the miter latch can be re-engaged at these angles. If there is no such notch, the arm can be clamped at any angle.

   d. Re-engage the miter latch and tighten the arm clamp.

   e. Make the cut in the same manner as described for crosscutting.

5. Bevel cuts:

   a. Select a crosscut or combination blade.

   b. Lock the radial arm and the motor yoke in the same position as for crosscutting.

   c. Raise the saw until the motor can be tilted to the desired bevel. Release the bevel clamp and the locating pin.

   d. Tilt the saw end of the motor downward to the desired bevel as indicated on the bevel scale.

   e. Re-engage the locating pin and tighten the bevel clamp.

   f. Make the cut in the same manner as described for crosscutting.

   g. Bevel rips can be made in a similar manner except the saw yoke is turned and locked in place as in ripping. Follow same procedures as discussed in straight ripping.

   h. A bevel-miter (compound angle cut) is a combination bevel and miter cut.

C. GENERAL SAFETY PRACTICES:

1. Wear eye protection, hearing protection, and proper clothing at all times when operating this machine.

2. Do not operate saw without permission from the instructor.

3. Be sure blade is sharp, sound, and of the proper type.

4. All adjustments should be tight and all guards in place.

5. Never leave tools, scraps, or other materials on saw table. Keep area around saw clear.

6. Do not leave machine while it is running.

7. Be sure machine is grounded properly.

8. Be sure material is free of knots, nails, or other foreign matter.

9. Do not adjust machine while it is running.

10. Tighten rip lock before starting the saw.

11. Pull saw slowly through material. Return saw to rear of table after sawing and before removing stock.

12. When ripping, make sure that the blade is rotating upward toward the operator. Feed the stock from the end opposite the anti-kickback device. Always use a push stick when ripping.

13. Do not stop blade by pushing stock against the blade.

14. Do not saw material freehand without a guide.

D. COMPLETION QUESTIONS:

1. _____ or _____ types of blades my be used for crosscutting.

2. A _____ blade or feeding the saw to _____ may result in the saw grabbing the material.

3. The depth of cut into the table is adjusted by turning the _____ _____.

4. Material should be fed opposite from the _____ when ripping.

5. The saw blade should be placed on the arbor so the teeth on the _____ of the blade point back toward the elevating column.

6. Be sure all _____ are tight before the saw is turned on.

7. Return the saw to the _____ of the table after making a crosscut.

8. A _____ angle cut is a combination bevel and miter cut.

9. When the saw is perpendicular to the table, the bevel gage should read _____ degrees.

10. To crosscut a board at other than a 90° angle across the board, the _____ must be released.

PORTABLE CIRCULAR SAW

A. PART IDENTIFICATION:

Identify the circled parts on the portable circular saw illustrated below.

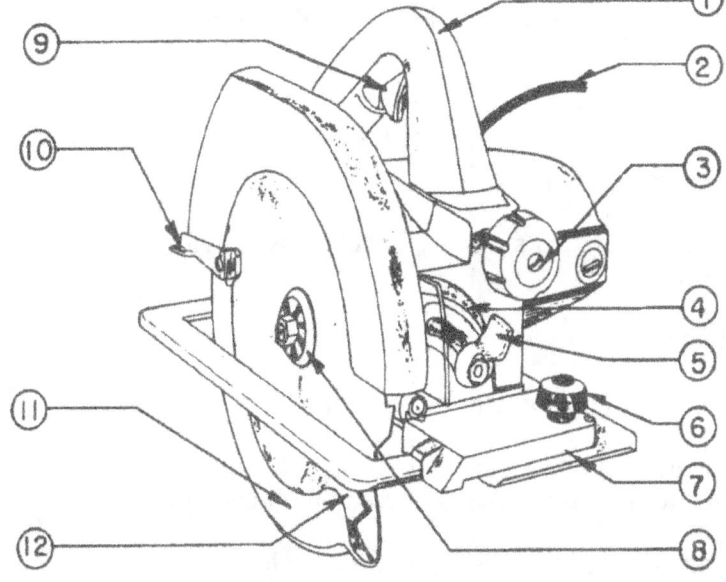

1. _____
2. _____
3. _____
4. _____
5. _____
6. _____
7. _____
8. _____
9. _____
10. _____
11. _____
12. _____

B. SAFE OPERATIONAL PROCEDURES:

1. Crosscutting

    a. Use crosscut or combination blade. Do not change blades or make any adjustment while saw is connected to electrical source.

    b. Be sure work is solidly supported. Do not cut materials between two supports, but allow material to fall or be lifted away from saw.

    c. Adjust the depth of cut so that the teeth clear the thickness of material by the depth of the teeth.

    d. Grasp handle firmly, do not place finger on starting trigger until ready to start saw. Place front of saw base on work so that the guide mark and line of cut are in line.

    e. Advance saw into wood, following line of cut. Save the full cutting line.

    f. Guide the saw steadily through the cut; if saw stalls, <u>do not</u> release starting trigger, but back saw out of check if saw is binding from attempting to wedge or turn in cut.

    g. When end of cut is reached, release trigger switch and allow the blade to follow through as the saw is lifted away from the work. Allow saw to stop before setting down.

2. Ripping:

   a. Use combination or ripping blade.

   b. Insert rip guide in frame if ripping a width that the guide will reach. For cuts wider than the scope of the guide, a straight board clamped to the workpiece can serve as a guide.

   c. If saw kerf seems too tight and binds the blade, a wedge can be inserted to open the kerf and give clearance to the blade. Follow steps b through g as in crosscutting.

3. Bevel and Miter cuts:

   a. Use combination or crosscut blade.

   b. Lay out angle or line of cut with a sliding T-bevel or use angle gage provided with most saws.

   c. Adjust saw to bevel angle.

   d. Adjust blade depth so that blade penetrates bevel thickness of material.

   e. The miter cut is made at an angle other than 90° across the board. The saw blade is set at 90° to the saw frame or zero on the bevel angle scale.

   f. A compound bevel cut is a miter cut using a bevel angle.

   g. Follow rules as in crosscutting.

4. Pocket cut:

   a. Select a combination blade.

   b. Mark area to be cut.

   c. Adjust depth of cut.

   d. Push telescoping guard forward so lower edge of blade is exposed.

   e. Starting near a corner limit of the pocket to be cut, tilt the saw forward on front of base until the front edge of the blade rests on the surface on the waste side of the line of cut.

   f. With the blade clear of the material, start the saw, lower slowly into the material until the base rests firmly on workpiece.

C. GENERAL SAFETY PRACTICES:

1. Wear eye protection and proper clothing.

2. Do not operate saw without instructor's permission.

3. Use only sharp blades for job to be done.

4. Double check all adjustment thumbscrews to be certain they are tight and locked at zero or proper adjustment needed for the job.

5. Be sure the saw is disconnected from power source when changing blades or making any adjustments.

6. Never start saw when blade is in contact with material being cut.

7. Saw should always be allowed to stop before setting on floor or bench.

8. Saw only in forward direction, <u>never</u> attempt to saw in reverse.

9. Keep right arm in line with blade, keep other arm and parts of body well out of danger.

10. Do not use saw in awkward position such a above head, on a ladder, or on sloping surfaces.

11. Be sure saw is properly grounded electrically. Determine location of cord at all times to avoid cuttings. Note: Power tools having double insulated or plastic handles and motor housings do not need to be grounded electrically and do not have 3-prong cord plugs.

12. Be sure telescoping gage works freely.

13. Do not reach over or around saw when running.

14. Allow saw to reach maximum speed before advancing into materials; do not force or attempt to turn saw in cut.

15. Do not talk to anyone while operating the saw.

16. Always check the location of the support material to avoid sawing into it.

D. COMPLETION QUESTIONS:

1. In making a long rip cut, it may be necessary to insert a _____ in the _____ to keep blade from binding.

2. A _____ or _____ blade can be used for crosscut sawing.

3. A _____ saw blade should not be used.

4. A _____ clutch, a safety feature on some portable circular saws protects the operator from possible kickback of the saw.

5. The _____ guard must be lifted before making a pocket cut.

6. A _____ bevel miter cut is the cutting of two angles at one time.

7. A portable circular saw for general shop work should be selected in the _____ to _____ inch blade size range.

8. The saw should be properly _____ to avoid electrical shock.

9. The thickness of the bevel will be _____ than the vertical thickness.

10. A _____ guide can be used when ripping to assure uniform width of the completed materials.

RECIPROCATING SAW

A. PART IDENTIFICATION:

Identify the circled parts on the reciprocating saw illustrated below.

1. _____
2. _____
3. _____
4. _____
5. _____
6. _____
7. _____
8. _____
9. _____
10. _____

B. SAFE OPERATIONAL PROCEDURES:

1. The reciprocating saw is designed to make cuts in any direction thus making it more versatile than the sabre saw. The blade moves in a straight line with the saw which makes for a simpler design and a better power factor than the right angle movement of the blade in the sabre saw.

2. The reciprocating saw can be used by many and varied construction workers such as carpenters, electricians, and plumbers, and has blades available for cutting wood, steel, plastic, and other materials.

3. Most reciprocating saws have two speeds, a high speed (approximately 3000 RPM) for wood and a low speed (approximately 2000 RPM) for metal cutting.

4. Generally the saw is designed to have two blade positions, a vertical and a horizontal. The blade position is adjusted by changing the blade position in respect to the blade shoe.

5. The length of stroke of the blade is usually 1/2" to 1".

6. Some reciprocating saws have a rocker type shoe which helps to roll the blade into starting cuts.

7. Blades for the reciprocating saw vary greatly in sizes, shapes, and types. Blades vary in length from 2-1/2 to 18 inches and are made of high speed steel for metal cutting and carbon steel for cutting wood and other soft materials. The common number of teeth per inch is 6-10 for wood and 18-24 for metal cutting.

8. Whatever the job to be done, the blade should be sharp, of proper type and correctly attached to the machine.

9. Making a plunge cut:

   a. Rest the saw on its shoe with the blade in line with the mark of the opening to be cut, yet not touching the wood.

   b. Pull the trigger switch allowing the blade to enter the wood. Gradually raise the angle of the saw until the blade enters the wood.

   c. The saw may then be held at any angle that is convenient for the operator.

   d. When a corner is reached, back up the saw and make several cuts to the corner line.

   e. As soon as enough material has been removed, turn the saw and cut along the new line.

   f. A curved plunged cut is similar except the starting cut should be made to the inside of the outline so as not to make the curved edge oversize. Also the saw must be held at a right angle to the material when cutting.

10. Making a straight or curved cut:

    a. In straight line cutting, set the shoe of the saw on the work.

    b. Start the motor and move the saw into the work but do not force the cutting action.

    c. Keep the shoe against the work at all times; the blade cuts from the bottom side to the top side, just opposite the cutting action of a handsaw.

    d. The good or finish side should be placed face down during the cutting operation.

    e. As with the curved plunge cut, when sawing curves the saw must be held at a right angle to the work.

11. A flush cut (cut close up against a wall) can be made with most reciprocating saws by simply adjusting the blade carrier (horizontal) so it will clear the body of the saw. Follow similar procedures as in making opening cuts.

12. The reciprocating saw can be used for cutting metal by using the correct blade and following the same general procedures as for sawing wood. One important point is to be certain the metal is secured, either as constructed or held in a vise.

13. The saw should be kept clean, well lubricated, and periodically checked for loose parts and screws.

14. The blade should be removed when storing the saw.

C. **GENERAL SAFETY PRACTICES:**

   1. Wear eye protection when operating the reciprocating saw.

   2. Obtain permission from the instructor to operate the reciprocating saw.

3. If it is not double insulated, make sure the saw is properly grounded.

4. Do not operate with loose fitting clothing.

5. Make sure the shoe is always against the work piece.

6. Always use sharp and proper type and size blade for the job being done.

7. Do not force the cut or attempt to turn too short a corner.

8. If using in an awkward position, such as above the head or on a ladder, make sure operator and the material being cut are properly and securely supported.

D. COMPLETION QUESTIONS:

1. The reciprocating saw is more versatile than the _____ saw because it cuts in a _____ _____ rather than at right angles to the saw base.

2. The high speed range is approximately _____ RPM and is recommended for cutting _____.

3. Metal cutting blades are made of _____ _____ steel and generally have _____ to _____ teeth per inch.

4. The finish side of the stock should be face _____ during the cutting operation.

5. The saw must be held at a _____ angle in making _____ cuts.

6. The _____ _____ helps to roll the blade into the cut.

7. Most saws can be adjusted for two blade positions, these being _____ and _____.

8. The _____ position would be used when making a flush cut.

9. The sawing action is from the _____ to the _____ of the stock.

10. The wood cutting blade would commonly have _____ to _____ teeth per inch.

NOTES

POWER MITER SAW

A. PART IDENTIFICATION:

Identify the circled parts on the power miter saw illustrated below:

1. _____
2. _____
3. _____
4. _____
5. _____
6. _____
7. _____
8. _____
9. _____
10. _____
11. _____
12. _____

B. SAFE OPERATIONAL PROCEDURES:

1. Study the operation, maintenance, and safety manual(s) for the specific saw to be operated.

2. Changing the saw blade:

   a. Disconnect saw from the power source.

   b. Select a crosscut or combination blade.

   c. Remove the saw guard, the arbor nut, and arbor collar. Remember the arbor has left-hand threads. Remove the blade.

   d. Place the blade on the arbor so the teeth toward the operator point downward.

   e. Place the outside collar on the arbor. Be sure the recessed face of both collars are against the saw blade.

   f. Tighten the arbor nut using the wrench furnished with the saw.

   g. Replace the guard. If it is the retractable type guard, make sure it moves freely before reconnecting the power source.

3. Miter cuts:

   a. Select a sharp crosscut or combination blade.

   b. The saw should be in the raised position and resting on the spring.

   c. Select the proper angle of cut by moving the spring loaded miter arm. Most saws have a stop at 90 degrees and 45 degrees left and right. Lock the miter arm in position for the desired cut with the lock nut.

   d. Place the stock in the saw with the flat or square edges against the table and the fence.

   e. Hold the material against the fence with one hand with the mark under the saw blade. Fingers or hands should never be within three inches of the path of the blade. Use a C-clamp to hold small pieces. Sand paper glued to the fence will prevent the stock from shifting.

   f. Start the motor before making contact with the material. Slowly lower the saw into the stock with hand positioned on the saw handle.

   g. After completing the cut, return the saw to the raised position. Allow the motor to stop before removing stock from the table.

4. Compound angle cuts:

   a. To cut a compound angle, first prepare a filler block to attach to the fence with screws.

   b. The angle at the front of the filler block will determine the bevel angle and the tool control setting will determine the miter angle.

C. GENERAL SAFETY PRACTICES:

1. Wear industrial quality eye protection and proper clothing when operating this saw.

2. Obtain permission from the instructor before operating the power miter saw.

3. Be sure the blade guard is in place and working properly.

4. Operate the miter saw only where adequate light is available.

5. Be sure the stock is firmly supported. Do not attempt to hold the stock away from the fence.

6. Always keep fingers more than 3 inches from the path of the blade.

7. The stock must be in contact with the fence near the blade to prevent pinching the blade.

8. Clean all scrap material and sawdust away from the work area before starting the saw.

9. Do not leave the work area until the saw blade has stopped.

10. When the job is completed, clean the saw and work area.

D.   COMPLETION QUESTIONS:

1. The _____ must be removed to change the blade.

2. The arbor nut has _____ hand threads.

3. The teeth of the blade near the operator are pointed in a _____ position.

4. After use, the spring returns the saw to the _____ position.

5. When making a cut, the saw should be lowered _____ into the workpiece.

6. The motor should _____ before removing stock from the table or leaving the work area.

7. To cut a compound angle, a _____ block must be used.

8. Fingers should never be closer than _____ inches to the path of the blade.

9. A _____ or _____ blade should be used for straight and miter cuts.

10. The _____ _____ is adjusted across the index or degree scale for making angle cuts.

NOTES

FINISHING SANDER

A. PART IDENTIFICATION:

Identify the circled parts on the finishing sander illustrated below.

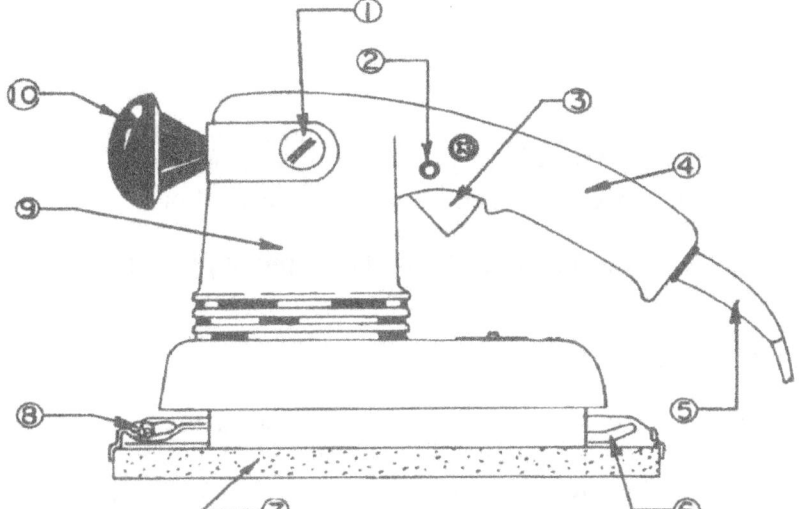

1. _____
2. _____
3. _____
4. _____
5. _____
6. _____
7. _____
8. _____
9. _____
10. _____

B. SAFE OPERATIONAL PROCEDURES:

1. The finishing sander is designed to do smoothing or fine sanding before the final hand sanding operation.

2. It may have an orbital (elliptical) stroke or a straight-line (straight) stroke.

3. The finishing sander is designed to use 1/3 or 1/2 of a standard 9" x 11" sheet of sandpaper.

4. Select the proper grit (coarseness or fineness) of sandpaper according to the following table indicating class, mesh size, and symbol of sandpaper.

CLASS	MESH SIZE	SYMBOL	CLASS	MESH SIZE	SYMBOL
	400	10/0		100	2/0
	360	———	Medium	80	1/0
Very	320	9/0		60	1/2
Fine	280	8/0		50	1
	240	7/0	Coarse	40	1 1/2
	220	6/0		36	2
	180	5/0	Very	30	2 1/2
Fine	150	4/0	Coarse	24	3
	120	3/0		20	3 1/2

5. Attaching sandpaper to machine:

   a. Make sure power is disconnected.

   b. Check condition of sandpaper for rips, tears, or paper being filled with glue or sanding dust.

   c. Select proper grade (coarseness) of sandpaper for job. Cut, do not tear, the sanding sheets to size.

   d. Open sandpaper clamps.

   e. Insert paper into front clamp, release clamp, then insert paper into rear clamp and release clamp. Make sure paper is straight, covers total pad, and has sufficient tension so it is tight and snug against pad.

6. Check off-on switch before connecting cord. If cord has 3-prong plug, make sure it is properly grounded.

7. Clamp stock to be sanded in vise or to the table.

8. Start the sander above the work. Place sander on workpiece surface, beginning at one side and moving the sander across the piece using only enough pressure (normal weight of the machine) to keep the sander cutting. Sand with the wood grain. Sanding is generally done with one hand. Use successively finer grit paper until desired finish is obtained.

9. When sanding is completed, lift the sander off the work before stopping the motor. Set sander on center of table and on a piece of hardwood so sander will not fall or damage table top.

10. Before storing machine, remove sandpaper and thoroughly clean dust and other material from machine.

C. GENERAL SAFETY PRACTICES:

1. Wear eye protection at all times when operating this sander.

2. Obtain permission from the instructor to operate the finishing sander.

3. Never operate machine with torn or worn sandpaper.

4. Make sure sander switch is off before connecting to power source.

5. Sander should be grounded electrically.

6. Check clamping devices to see that paper is secure before operating machine.

7. Do not overload machine with excessive pressure down on workpiece.

8. Keep machine clean and well lubricated.

D. **COMPLETION QUESTIONS:**

1. Two types of finishing sanders as far as types of strokes are the _____ and _____ types.

2. The finishing sander usually takes _____ or _____ of a sheet of 9" x 11" sandpaper.

3. A very fine sandpaper having a symbol of 9/0 would be an abrasive mesh size of _____.

4. A number 80 mesh size would be considered _____ class sandpaper.

5. Downward pressure on the workpiece is generally enough just from the _____ of the machine.

6. The paper should be inserted into the _____ clamp and then stretched over the _____ and locked into the _____ clamp.

7. Sandpaper should be _____ rather than _____ to size.

8. The machine should be moved _____ the wood grain.

9. The sanding operation is normally completed by holding the machine with _____ hand.

10. The use of torn paper could result in damage to the sander _____.

NOTES

BELT SANDER

A. PART IDENTIFICATION:

Identify the circled parts on the belt sander illustrated below.

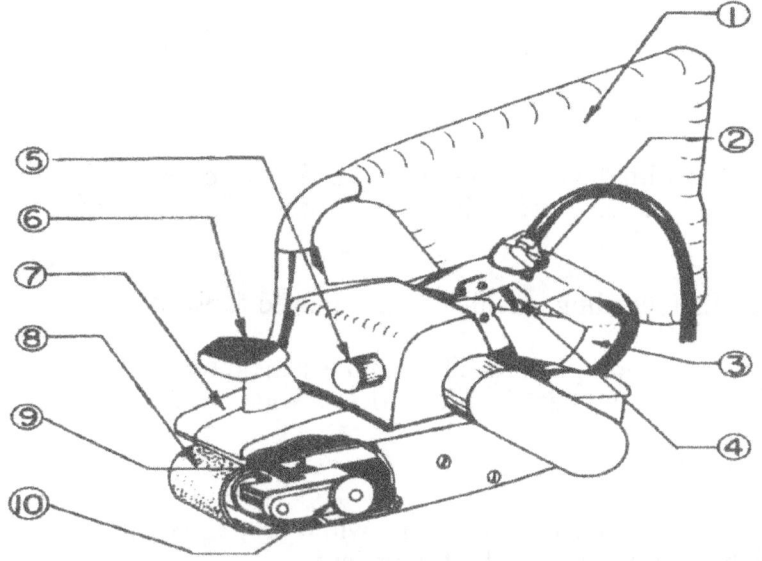

1. _____
2. _____
3. _____
4. _____
5. _____
6. _____
7. _____
8. _____
9. _____
10. _____

B. SAFE OPERATIONAL PROCEDURES:

1. The belt sander is sized by the width of the sanding belt; a 3-inch width is the most common size.

2. The belt turns on two pulleys or spindles, one of which is adjustable for tightening and aligning the belt.

3. Select the correct size belt for the specific machine and coarseness according to the job to be sanded. (Note class of sandpapers in finishing sander lesson.)

4. Installing the belt:

    a. Make sure power is disconnected before removing or replacing the belt.

    b. Check condition of belt to see that it is not frayed on the edges, torn, or filled with glue and sanding dust.

    c. Loosen or release belt tightening mechanism.

    d. Place belt on pulleys-note arrow on back of belt for proper direction of travel.

    e. Tighten belt so it has tension keeping it snug against the pulleys and sander base.

    f. Check the off-on switch to see that it is "off" and not locked in the "on" position.

g. Connect cord to power source. If cord has 3-prong plug, make sure it is properly grounded.

h. With sander on its side so that the belt will not touch table or other materials, or with sander upside-down and well supported, pull trigger switch and check belt alignment or tracking. Turn tracking adjustment knob to make sure the belt is tracking on the center of the front pulley. If off center, the belt may cut through the sander housing.

5. Before starting sanding operation, check the dust bag and empty if it contains dust.

6. Clamp stock to be sanded in vise or to table.

7. Hold the sander in both hands with the right hand on the switch or back handle and the left hand on the front handle. Start sander above the work, at one end or side of workpiece.

8. Allow rear of belt to touch first, leveling machine as it is moved forward.

9. Move the sander lightly over the workpiece, sanding in the same direction or with the grain, and moving the sander back and forth over a wide area.

10. Raise the sander slightly at the end of the stroke.

11. Begin the second stroke by lapping halfway across the width of the first stroke and continuing this procedure across the width of the board.

12. Use progressively finer grit belts until the desired finish is obtained.

13. Do not pause in any one spot, but keep the sander moving over the workpiece at all times.

14. Lift the sander off the work before allowing switch to go to the "off" position.

15. Wait until the motor is completely stopped before placing the sander on bench.

16. When the work is completed, disconnect the sander from the power source, empty the dust bag, blow off belt and machine with compressed air, and place sander in proper storage area.

C. **GENERAL SAFETY PRACTICES:**

1. Wear safety glasses at all times while operating this sander.

2. Obtain permission from instructor to operate belt sander.

3. Wear proper clothing.

4. Never use sanding belt in poor condition.

5. Check motor switch to see that it is off before connecting to power source.

6. Check alignment or tracking and tension of belt.

7. Do not attempt to take material off too fast by excessive pressure and over-working the motor.

8. Unplug sander when not in use.

9. When not sanding, set sander on piece of hardboard near center of bench where it will not fall or damage the bench top.

10. Keep the sander clean and well lubricated.

D. **COMPLETION QUESTIONS:**

1. Belt sanders are sized by the _____ of the belt with a common size being _____ inches.

2. Before connecting the sander, make sure the switch is in the _____ position.

3. The belt is adjusted back and forth on the front pulley by turning the _____ knob.

4. In starting the sanding operation, set the _____ of the sander down first.

5. Always sand in the same direction of the _____, moving the sander _____ and _____ over a wide area.

6. In finishing a job, use progressively _____ _____ sanding belts until the desired finish is obtained.

7. When not sanding, set the sander on a piece of _____ and near the _____ of the bench.

8. Keep the machine _____ and well _____ for long life.

9. Never set the sander down until the motor is completely _____.

10. Place the belt on the pulley with the arrow _____ the direction of travel.

NOTES

DISC SANDER

A. **PART IDENTIFICATION:**

Identify the circled parts on the disc sander illustrated below.

1. _____
2. _____
3. _____
4. _____
5. _____
6. _____
7. _____
8. _____

B. **SAFE OPERATIONAL PROCEDURES:**

1. The disc sander is useful for sanding or shaping edges and end grain of stock.

2. The disc sander is sized by the diameter of the disc in inches, with 12 inches being a common size.

3. Select the coarseness of the abrasive disc according to the more common sanding jobs to be completed. Keep the sandpaper clean and in good condition.

4. The sanding pad is attached to the metal disc with special adhesive that will securely hold the sandpaper to the disc, yet allow it to be removed when it becomes smooth or filled with sanding dust.

5. The disc sander is usually equipped with an adjustable table on which a miter gage is used to guide the stock when sanding bevels and angles. The table can be adjusted up and down to sand bevels on the edge of stock. A square can be used to check the angle between the table and the disc.

6. Adjust the table so edge is 1/8" or less from the disc. Make all adjustments with motor power off and disc stopped.

7. Keep the disc guard in place at all times.

8. To sand or smooth the end of a board, place the board on the sander table so the disc cuts in a downward direction on the board.

9. Turn the motor on, allow the motor to gain full speed, and then move the workpiece carefully into the disc with enough pressure to keep the disc cutting.

10. Move the stock sideways (back and forth) slightly to reduce the heat caused by friction between the disc and the edge of the stock.

11. When the stock is smooth or to the desired sanding line, remove it from the table and turn off the motor.

12. Do not leave the safety zone area until the motor has completely stopped.

C. GENERAL SAFETY PRACTICES:

1. Wear eye protection at all times during operation of this machine.

2. Obtain permission from the instructor to operate the disc sander.

3. Never operate the machine if the sandpaper is loose, torn, or filled with sanding dust.

4. Make all adjustments with the motor off and the disc completely stopped.

5. Do not allow hands or fingers to get near or touch the revolving disc.

6. Always sand on the side of the disc that is moving down toward the table.

7. Keep table edge within 1/8" of the disc.

8. Do not overload the motor with excessive pressure.

9. The disc sander is designed to smooth edges or end grain of stock and not for cutting excessive amounts from edges or ends of boards. Long periods of continuous sanding will overheat the disc, causing the stock to be discolored and possibly damaging the machine.

10. Make sure the machine is grounded electrically.

11. Do not talk to others while operating the machine.

12. Keep all guards in place, and the machine clean and well lubricated.

D. COMPLETION QUESTIONS:

1. The disc sander is sized by the _____ of the disc with a common size being _____ inches.

2. The _____ can be tilted for sanding bevel edges.

3. The abrasive pad is attached to the disc with a special _____.

4. The disc sander is designed for sanding _____ and _____ grain of stock.

5. Sanding should be done on the side of the disc that is moving _____ the table.

6. The edge of the tilting table should never be more than _____ inch from the disc.

7. The _____ _____ should be used when sanding mitered edges.

8. Continuous sanding in one location on the disc could cause the disc to _____.

9. A _____ could be used to check the angle between the table top and sanding disc.

10. Allow the disc to gain full _____ before beginning the sanding operation.

NOTES

WOOD LATHE

A. PART IDENTIFICATION:

Identify the circled parts on the wood lathe illustrated below.

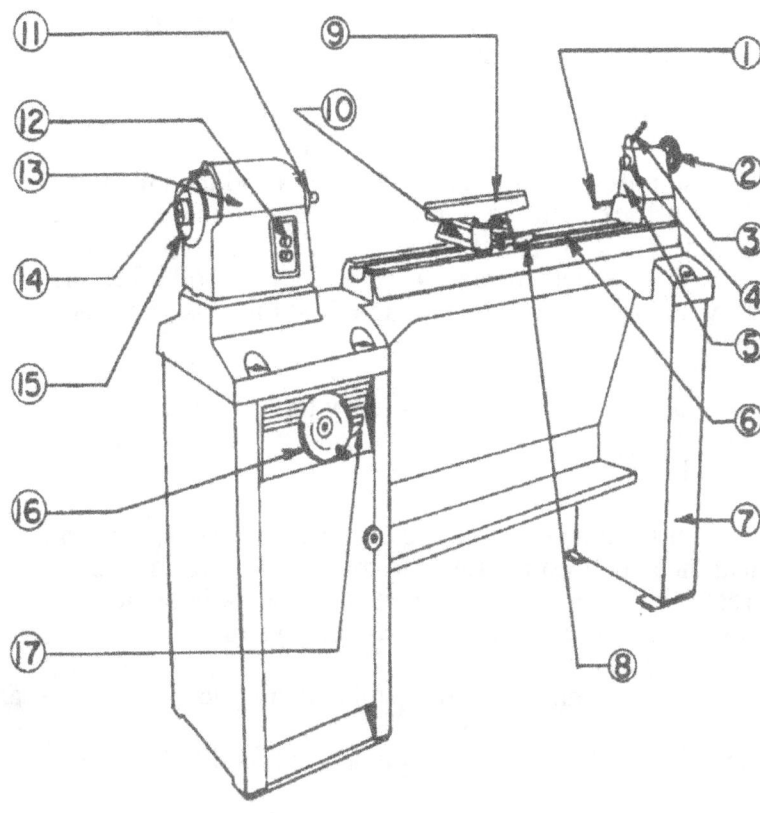

1. _____
2. _____
3. _____
4. _____
5. _____
6. _____
7. _____
8. _____
9. _____
10. _____
11. _____
12. _____
13. _____
14. _____
15. _____
16. _____
17. _____

B. SAFE OPERATIONAL PROCEDURES:

1. The wood lathe should be considered one of the more dangerous shop power tools due to the fact the fast moving material or workpiece cannot be fully guarded.

2. The speed of operation of the lathe is of particular importance as wood materials can actually blow apart due to running too fast. All wood lathes should have a method of changing the speed of operation. Always operate at prescribed speeds for the job.

3. Spindle turning or between-center turning:

    a. In spindle turning or between-center turning, the stock is mounted between the lathe centers.

    b. Select a piece of stock approximately one inch longer than the completed job. It should be free of loose knots, checks, cross-grain, and other wood defects.

c. If larger than 2 inches square, saw or plane off the corners.

d. Square the ends and find the center of each end by drawing diagonal intersecting lines.

e. Select one end as the live center and cut saw kerfs about 1/8" deep along the diagonal lines. Drive the spur (headstock) center into the kerfs with a mallet.

f. On the other end, at the point where the lines intersect, center punch a hole approximately 1/8" in diameter and 3/16" to 1/4" deep.

g. Replace the live center spur holding the workpiece into the headstock.

h. Adjust the tail stock spindle so that it advances 1" beyond the tail stock housing. Slide the tail stock until the point of the dead center enters the hole in the workpiece. Lock the tail stock in place. Turn the tail stock spindle feed handle until the dead center is seated in the wood. Release the pressure slightly and apply oil wax or soap to the impression in the end of the workpiece.

i. Turn the feed handle until the dead center is fairly tight in the original position. Back off slightly until the stock rotates freely on the dead center. Lock the spindle in place.

j. Adjust the tool rest until it is 1/8" above the center of the workpiece and there is a 1/8" clearance between the rest and the widest portion of the stock. The tool rest should be moved forward as the stock is reduced in size with a clearance of not more than 3/8 of an inch at any time.

k. Select a gouge tool for the rough work. Remember cutting tools must be sharp.

l. For the rough work, the machine should run at a low speed (600 RPM).

m. Assume a natural position with the feet slightly spread with one foot slightly ahead of the other.

n. Grasp the handle of the tool well toward the end with the right hand. Hold the blade with the left hand guiding it along the tool rest. Work from the center toward the end.

o. Continue turning until the workpiece is cylindrical and near the finished size. Finish the turning job with a skew cutter. The speed may be increased to 1600-1800 RPM for smoother cutting.

p. Depending on the completed job desired, select other cutters accordingly.

q. When the job is complete with the cutters, remove the tool rest for sanding. Select proper sandpaper and hold a long piece between both hands and at right angles to the stock with the right hand over the work. Bear down evenly, moving the paper back and forth on the work.

4. Headstock or faceplate turning:

a. In headstock or faceplate turning, the stock is mounted to a flat metal plate which is attached directly to the headstock.

b. Band saw a round disc at least 1/8" larger than finished dimension.

c. Attach a faceplate smaller than the stock to the stock with No. 12 wood screws. Make sure screws are sufficient length to hold the stock, yet not too long as to get hit by the cutting tool.

d. Remove the spur center and screw the faceplate and stock in the headstock spindle.

e. Adjust the tool rest parallel to the center of the stock and about 1/8" from the face of the stock.

f. Select the slowest speed (600 RPM) on the lathe. Turn the stock by hand to see that it clears the tool rest.

g. Rough turn the outside of the stock with a gouge, and then smooth with a round nose tool.

h. Reset the tool rest so it is parallel to the center surface of the stock. Using a round nose cutter, remove the inside or center of the stock. Continue turning until the bottom and walls are to the desired thickness, never less than 3/8" on the bottom or 1/4" on the wall thickness.

i. For sanding, remove the tool rest and fold a medium or fine sandpaper into a pad for smoothing the workpiece. Use a lathe speed of 1500-1800 RPM.

j. If turning laminated, glued stock, make sure the glue joints are secure. Never turn the lathe to its top speed on laminated stock.

C GENERAL SAFETY PRACTICES:

1. Wear eye protection and remove loose clothing or tie when operating the lathe.

2. Obtain instructor's permission to use the lathe.

3. Select stock for turning that is free of knots, poor graining, and splinters. Also, avoid turning stock that is improperly glued.

4. Check to see if stock is properly centered and secured firmly between centers.

5. Spin work by hand before turning on power.

6. Always use sharp tools and the proper tool for the job.

7. Make sure tool rest is approximately 1/8" from stock and no more than 3/8" away at any time.

8. Remove tool rest before sanding.

9. Do all rough cutting at low speed.

10. Avoid heavy cuts.

11. Hold cutting tool firmly and at proper angle to work. Don't leave tools on bed of lathe.

12. Stop the lathe often to check tail stock adjustment.

13. Keep tail stock (dead center) well lubricated.

14. Make all adjustments with power off.

15. Select proper and safe speeds for all operations.

16. Keep work area clean and free of chips and shavings at all times.

17. Never touch workpiece while it is turning no matter how smooth it may seem.

18. Do not talk to anyone while operating the lathe.

D. COMPLETION QUESTIONS:

1. All work turned in the lathe should be started with a _____ speed generally around _____ RPM.

2. Before starting the lathe, always turn the stock by hand to see that it clears the _____ _____.

3. To turn a small bowl, the stock is attached to the _____.

4. The center of the stock is determined by drawing _____ lines and the center is at the _____ of the two lines.

5. Most tail stock spindles remain solid and are called _____ centers.

6. The recommended speed for sanding is approximately _____ RPM.

7. When beginning a job, the tool rest should be set at _____ from the stock and never allowed to become more than _____ from the stock as the turning operation progresses.

8. A _____ is the cutting tool commonly used for rough cutting.

9. The tail stock should be _____ from time to time to keep it from burning or causing the stock to lock up in the lathe.

10. The _____ or headstock center is driven into the workpiece before inserting into the lathe.

PLANER-SURFACER

A. **PART IDENTIFICATION:**

Identify the circled parts on the planer-surfacer illustrated below.

1. _____
2. _____
3. _____
4. _____
5. _____
6. _____
7. _____
8. _____
9. _____
10. _____
11. _____
12. _____

B. **SAFE OPERATIONAL PROCEDURES:**

1. The planer-surfacer is designed to machine stock to exact thickness. It is equipped with a cutterhead, generally containing three knives, and it is similar to a jointer except it cuts the stock from the top.

2. The machine is sized by the width of the cutterhead with common sizes being 12, 18, and 24 inches.

3. It is equipped with four rollers, two upper and two lower. The upper rolls consist of the infeed roll, which is milled or corrugated to feed the stock into the machine, and the outfeed roller to pull the stock through the machine.

4. The chip breaker, located near the cutterhead on the infeed side, keeps the knives from chipping the board as they cut into the stock. The pressure bar, located near the head on the outfeed side, keeps the stock down against the table to get an even and uniform finished surface.

5. The depth of cut is determined by the location of the lower table bed in respect to the cutterhead, based on the starting thickness of the stock.

6. To plane a board to exact thickness, first determine the desired thickness and the starting thickness. Measure the board at its thickest point.

7. Set the planer-surfacer for the thickness of the board less the depth of cut. The recommended depth of cut for one pass is no more than 1/16 of an inch for hardwoods and no more than 1/8 of an inch for softwoods.

8. Position the stock so the knives will cut with the grain, with the first or true surface being down. Never plane or surface painted or varnished stock, or material containing nails or other foreign material.

9. Adjust the feed roll speed. The speed range is commonly 15 to 45 feet per minute. Softwoods may be run at the upper end of the range while hardwoods should be run at the lower end of the range.

10. Turn on the machine and allow the motor to gain full speed.

11. Feed the stock in at right angles to the cutterhead.

12. Do not force the material through the machine, but allow the feed roll to pull the stock through the machine.

13. If planing long material, get assistance from a helper or use support stand.

14. Be sure to stand to one side of the machine and never allow hands to get close to the feed rolls or knives.

15. Run the stock through as many times as necessary to reduce to the desired thickness. Successive cuts should be taken off alternate faces. Remember, the stock must be turned end for end so that planing will always be with the grain.

16. Pieces shorter than 12 inches should never be surfaced or planed.

17. Never plane or surface a board to less than 3/8 inch thick without the use of a backer board thicker and wider than the stock being planed.

18. Shut off machine and do not leave area until the machine has completely stopped.

C. GENERAL SAFETY PRACTICES:

1. Wear eye and hearing protection at all times when operating this machine.

2. Do not operate machine without permission from the instructor.

3. Make sure knives are sharp and properly adjusted.

4. Make all machine adjustments before turning the machine on.

5. Keep floor and work area free of chips, wood scraps, and other materials.

6. Make sure motor is at full RPM before feeding material into machine.

7. Always stand to one side when feeding or receiving stock.

8. Never attempt to remove more than 1/8" on softwoods or 1/16" on hardwoods during one pass.

9. Do not place hands near feed rolls or knives.

10. If stock gets stuck, shut machine off, lower the table bed, and remove the stock.

11. Remove loose knots, nails, or other defects before planing.

12. Do not plane stock shorter than 12 inches.

13. Do not plane stock to a thickness of less than 3/8" without a backer board.

14. Never talk to fellow workers while operating the machine.

15. Do not wear loose fitting clothes when operating the machine.

16. If board is wet, lubricate the bed with kerosene or wax.

D. **COMPLETION QUESTIONS:**

1. The planer-surfacer is sized by the _____ of the _____.

2. There are _____ rollers in the machine, _____ on top and _____ on the bottom.

3. The function of the _____ _____ is to keep the knives from chipping the board as it is being planed.

4. The maximum recommended cut per pass for softwoods is _____ of an inch.

5. The feed roll is _____ so that it can grip the board and push it through the machine.

6. Stock shorter than _____ should never be planed.

7. The feed roll speed on most planer-surfacers will range from _____ to _____ feet per minute.

8. The first run depth of cut is determined by the _____ part of the stock.

9. The stock should be fed into the machine at _____ angles to the cutterhead and so the knives are cutting _____ the grain.

10. Stock thinner than _____ of an inch should never be planed unless a _____ _____ is used.

NOTES

PORTABLE POWER PLANER

A. PART IDENTIFICATION:

Identify the circled parts on the portable power planer illustrated below.

1. _____
2. _____
3. _____
4. _____
5. _____
6. _____
7. _____
8. _____
9. _____
10. _____
11. _____
12. _____

B. SAFE OPERATIONAL PROCEDURES:

1. Study the operation, maintenance, and safety manual(s) for the specific power planer to be operated.

2. Keep the knives sharp and properly adjusted. Set screws are used to hold the knives in place. Adjust the knives so their cutting edges just contact a straight edge when it is placed against the rear shoe of the planer.

3. When knives are replaced, they should be sharpened and replaced as a set. Both knives should weigh the same to maintain balance of the cutter head.

4. The depth of cut is adjusted by raising or lowering the front shoe of the planer. Start with a very thin cut, 1/16" or less.

5. The fence should be used when planing the edge of a piece of lumber. Set the fence perpendicular to the bottom of the plane to get a square edge.

6. The fence must be removed when planing flat surfaces.

7. When starting a cut, apply downward pressure on the front edge shoe and move forward slowly. Control the plane forward movement, keeping it slow and uniform.

8. To finish a cut, put more downward pressure on the rear shoe as the planer is moved forward over the stock.

9. Plane with the grain, if possible, to avoid chipping.

C. **GENERAL SAFETY PRACTICES:**

1. Wear industrial quality eye protection and proper clothing when operating this machine.

2. Obtain permission from the instructor before operating the planer.

3. Be sure the chip deflector is properly attached to the machine so the chips will fly away from the operator.

4. Work on new wood. Nails and paint will damage the planer blades.

5. Be sure the work area is clean and free of scraps, sawdust, or chips.

6. Be sure the lumber is properly secured in a vise so it will not slip or tilt when downward and forward pressure is placed on the stock being planed.

7. Disconnect the planer from the power source when replacing or adjusting the blades or making any adjustments to the planer.

8. Do not make depth of cut adjustments while the motor is running.

9. Never set the plane down until the motor stops turnning.

10. Do not talk to anyone while using the planer. Concentrate on the work.

11. Hold the machine with one hand on the handle and the other on the knob, if one is available. Be careful not to come in contact with the knives after the cut is finished.

12. Be sure the planer is electrically grounded or double insulated.

13. Keep the power cord away from the knives.

14. After use, disconnect the power cord, clean the planer, and return it to proper storage.

D. **COMPLETION QUESTIONS:**

1. To make a uniform cut at the start of the stock, apply the most pressure on the _____ of the plane.

2. When the knives are replaced, they must be adjusted by using a _____ _____ along the knife edges and the rear shoe.

3. The depth of cut is adjusted by raising or lowering the _____ _____ of the planer.

4. The _____ _____ prevents the chips from flying into the operator's face.

5. The _____ _____ keeps the cord trailing to the left or right to help avoid getting it into the knives.

6. The fence is used to help make a _____ cut on the edge of the board.

7. Plane _____ the grain when possible to avoid chipping the grain.

8. The depth of cut adjustment must be made while the motor is _____.

9. All knives should weigh the same to maintain _____ of the cutter head.

10. _____ screws are used to hold the knives in place.

NOTES

SHAPER

A. PART IDENTIFICATION:

Identify the circled parts on the shaper illustrated below.

1. _____
2. _____
3. _____
4. _____
5. _____
6. _____
7. _____
8. _____
9. _____
10. _____

B. SAFE OPERATIONAL PROCEDURES:

1. Study the operation, maintenance, and safety manual(s) for the shaper to be operated.

2. Changing the shaper cutters:

    a. Disconnect the shaper from the power source or lock the switch in the off position.

    b. Select the proper cutter for the job; be sure it is sharp.

    c. Have the proper tools available for loosening and tightening the spindle nut.

    d. Lock the spindle so the nut can be loosened and tightened (check the manufacturer's procedure).

    e. Insert the cutter so the sharp edge will rotate into the wood.

    f. Best results are usually obtained by placing flat material upside down on the table with the cutter on the underside. This procedure prevents damage to the stock if it raises from the table, produces a better job if the material is uneven in thickness, and makes the operation safer by covering the cutter with the stock.

3. Cutting a rabbet, dado, or molding:

   a. Select the correct cutter for the job. Place the proper collars above and below the cutter. For most jobs, the collar is used to control the depth of the cut.

   b. Secure the cutter to the spindle by tightening the nut.

   c. Adjust the height of the spindle and lock in place.

   d. Adjust the fence so it is close to the cutter. The fence partially shields the cutter and spindle from the operator.

   e. The fence can be used to control the depth of the cut on straight stock. Be sure the fence clears the cutters. At least half of the stock must be against the fence at all times.

   f. Turn on motor and make a practice cut on a piece of scrap stock.

   g. When making a cut, move the stock against the rotation of the cutter.

   h. Cut end grain first to avoid chipping finished corners.

   i. Keep the stock moving slowly but steadily to prevent it from burning.

   j. Avoid excess pressure against the collar when it is used for controlling depth of cut or burning at the collar may result. Be sure at least 1/8" of stock is against the collar.

   k. Keep hands away from the cutters. Never allow hands to get closer than 6" from the cutter. Use push blocks when working with small pieces.

   l. Turn off the motor after the cut is completed and wait for the motor to stop before leaving work area.

   m. Remove the cutter and collar from the shaper after the job is completed, and return them to their proper storage places.

   n. If the shaper has a reversible-rotating motor with a reversing switch, always use a keyed washer under the spindle nut to avoid having the spindle nut loosen during operation.

C. **GENERAL SAFETY PRACTICES:**

1. Wear industrial quality eye protection and remove all loose fitting clothing when operating this machine.

2. Obtain permission from the instructor before operating the shaper.

3. Use only sharp cutters; be sure they are mounted properly.

4. Be sure the shaper is disconnected from the power source or the off-on switch is locked in the off position when changing cutters or making any adjustments.

5. Never start the shaper when the cutter is in contact with the stock to be cut.

6. Do not talk to anyone while operating the shaper.

7. Keep the work area free of scrap material, but don't try to remove small pieces from the area near the cutter until it stops.

8. Be sure adequate light is available.

D.   COMPLETION QUESTIONS:

1. The cutter is held in place by a _____ threaded on to a short shaft called a _____.

2. The cutter must be mounted so its sharp edge will rotate _____ the stock.

3. The stock must be fed into the cutter _____ the rotation.

4. The _____ is used to control the depth of cut if the edges of the stock are not straight.

5. The _____ can be used to control the depth of cut on straight sided stock.

6. _____ grain should be cut first to avoid chipping the finished corners.

7. If the movement of the wood is stopped while in contact with the rotating cutter or collar, _____ of the stock will result.

8. When shaping flat stock, best results will usually be obtained if the collar is mounted _____ the cutter.

9. If the shaper has a reversible-rotating motor, a _____ _____ must be used under the spindle nut.

10. Never allow hands to get closer than _____ inches from the cutters.

NOTES

ROUTER

A. **PART IDENTIFICATION:**

Identify the circled parts on the router illustrated below.

1. _____
2. _____
3. _____
4. _____
5. _____
6. _____
7. _____
8. _____
9. _____
10. _____
11. _____
12. _____

B. **SAFE OPERATIONAL PROCEDURES:**

1. Changing the router bits or cutters:

    a. Disconnect router from power source.

    b. Select proper bit for job to be completed.

    c. Have proper router chuck and collet tools available for loosening and tightening chuck.

    d. Loosen locking handle and remove the router base.

    e. Check manufacturer's procedure for changing bits, in particular, for method of holding motor or chuck to properly tighten bit.

    f. Insert router bit at least 1/2" into chuck. Tighten securely. Turn router by hand to make sure the bit clears the router base. Replace router base.

    g. Release motor locking device before connecting router to power source.

2. Cutting a rabbet, dado, or molding:

    a. Select proper cutter for the job. It must be sharp. Replace plastic window over cutter area if router has a plastic window cover.

b. Clamp work securely and make all adjustments before starting the router.

c. Lock the cutter in the router and adjust the base to the desired height using depth adjustment gage. Lock depth adjustment.

d. If cutting a groove, a fence guide is required. Insert fence adjustment bars into router base and tighten lock screws.

e. Place the router base on the work with the cutter clear of the wood before turning on the power. Adjust fence guide.

f. Hold router firmly when turning on the motor to resist starting torque. A well-balanced stance is important to help maintain full control.

g. Make a practice cut on a piece of scrap lumber.

h. Hold the router with both hands, using the handles provided.

i. When making a cut along a straight edge, move left to right. Cut end grain first, then edge grain to avoid chipping of ends.

j. When making circular cuts, move counterclockwise around the circle.

k. Use steady, slow, even feed. Don't overload the motor.

l. Cut only clean lumber, free of paint, varnish, and nails. Keep in mind the cutting of plywood dulls the cutter because of the glue in the plywood.

m. Make sure the fence gage is tight against the edge.

n. When the cut is completed, turn off the motor. Do not lift the router from the work until the motor has stopped.

o. Between cuts lay the router on its side on a table in a position where it will not roll.

p. Remove the cutter from the router after the job is completed and return it to its proper storage place.

q. Special precaution should be used in operating this power tool. The cutter bit travels between 20,000 and 30,000 RPM and cannot have a guard or it would not function.

C. **GENERAL SAFETY PRACTICES:**

1. Wear eye protection and remove loose clothing when operating this machine.

2. Do not operate router without permission from the instructor.

3. Use only sharp cutters for the job to be done.

4. Double check all adjustments to be certain they are tight. Replace plastic window (if one is provided).

5. Be sure the router is disconnected from power source when changing cutters.

6. Never start the router when the cutter is in contact with the material.

7. Be sure the router is properly grounded electrically. Note location of cord to avoid cutting it. New models may be double-insulated; therefore, the cord will not have the three-prong grounded type electrical plug.

8. Do not talk to anyone while operating the router.

9. Check to see if the switch is off before inserting the plug into the outlet.

10. Work area must be free of obstructions such a scrap boards.

D. COMPLETION QUESTIONS:

1. The router is a dangerous power tool because the cutter bit travels at _____ RPM and the bit is not _____.

2. The router bit is held in a _____ type chuck.

3. The variety of router cuts is obtained from the great assortment of _____ available.

4. The bit should be inserted at least _____ into the chuck.

5. The commonly used freehand cuts are the _____ and _____ cuts.

6. The cutter will be dulled when cutting _____ because of the glue.

7. A router should be left lying on its _____ between cuts.

8. A _____ cutter bit should not be used.

9. Be sure switch is _____ before inserting the plug into an outlet.

10. In making long straight edge cuts, the direction of travel should be from _____ to _____.

NOTES

JOINTER

A. **PART IDENTIFICATION:**

Identify the circled parts on the jointer illustrated below.

1. _____
2. _____
3. _____
4. _____
5. _____
6. _____
7. _____
8. _____
9. _____
10. _____

B. **SAFE OPERATIONAL PROCEDURES:**

1. Keep the guard covering the cutterhead in place and free to function at all times.

2. Remember, the jointer's main function is to straighten the edge of lumber and it should not be used as a surfacer or planer.

3. Keep the knives sharp and properly adjusted. Dull knives are more likely to cause a kickback. Knives out of adjustment (either cutting on an angle or one knife adjusted lower or higher than the other) will cause improper cutting and possibly a safety hazard.

4. When knives are replaced, all knives should be replaced and be sharpened alike as a set. Due to the speed of the cutterhead (approximately 6000 RPM), knives must be in balance. Most cutterheads will have three knives.

5. The depth of cut is adjusted with the front table adjusting hand wheel. The machine should never be set to cut more than 1/8" for softwoods or 1/16" for hardwoods.

6. The depth of cut scale should be checked and readjusted to zero each time new blades are added or after adjustments to the cutterhead.

7. To do satisfactory work, the rear table must be exactly level with the knives in the cutterhead. Check by laying a straightedge over the table and cutterhead and then adjusting the table up or down as needed. Only a qualified person should make this adjustment.

8. The fence should be moved back and forth to different positions periodically so as not to cut at the same position on the cutterhead. Check the angle of the fence with a square and adjust fence pointer or scale to the zero point.

9. Bevel cuts may be made by adjusting the fence control handle to the proper angle. Bevel cuts at sharp angles are more dangerous as the stock will tend to slide down the fence and float over the cutterhead.

10. Check grain of lumber to be certain the knives are cutting with the grain and not against it. A beginner should not attempt to joint end grain.

11. Have a push-stick handy so it can be used on short pieces or at the end of a board so as to keep hands as far from cutterhead as possible.

12. If jointing long pieces, use a roller support or have someone support the piece after it passes over the rear outfeed table.

13. Do not stand directly in line with the tables in case of a kickback. Also see that others are not in line with the jointer when in use.

14. The cutterhead turns down toward the front infeed table; therefore, the revolving cutterhead will have a tendency to throw the work back toward the operator.

15. Do not attempt to joint stock less than 1/2 inch thick or pieces shorter than 12 inches in length.

16. Push stock into the jointer slowly or at about 16 feet per minute. Make certain the stock is snug against the table and fence surfaces as it is moved across the tables.

17. Turn off the power and allow machine to stop before leaving the area.

18. Test squareness of edges with a square and by stacking pieces of edged lumber on edges.

C. **GENERAL SAFETY PRACTICES:**

1. Wear eye protection and proper clothing when operating this machine.

2. Do not operate jointer without instructor's permission.

3. Use only sharp knives.

4. Keep floor and area around jointer clean and free of scraps and rubbish.

5. Do not make any adjustments with machine running.

6. Double check all adjustments before turning on power.

7. Do not attempt to joint or edge material containing nails, paint, severe knots, or other defects.

8. Always keep guard in place and working freely.

9. Do not reach over or around cutterhead while machine is running.

10. Talking to others while operating, or diverting attention to something other than the jointer, can cause accidents.

D. **COMPLETION QUESTIONS:**

1. Stock shorter than _____ inches should not be run on the jointer.

2. According to recommended speed of moving stock over the jointer, it would take approximately _____ seconds to edge an 8-foot board.

3. The depth of cut is adjusted by moving the _____ table up or down.

4. The dual fence control handle is used to adjust _____ angles and position of the _____ back and forth over the cutterhead.

5. Stock should be at least _____ thick to be run on the jointer.

6. A _____ _____ should be used for short stock and in finishing the end of a long stock as it passes over the cutterhead.

7. The _____ table should only be adjusted by a qualified person.

8. All _____ should be changed on the cutterhead when dull to keep it in _____.

9. The stock should be fed into the jointer so the machine is cutting _____ the grain and not _____ the grain.

10. The jointer should never be operated with the _____ removed.

NOTES

DRILL PRESS

A. PART IDENTIFICATION:

Identify the circled parts on the drill press illustrated below.

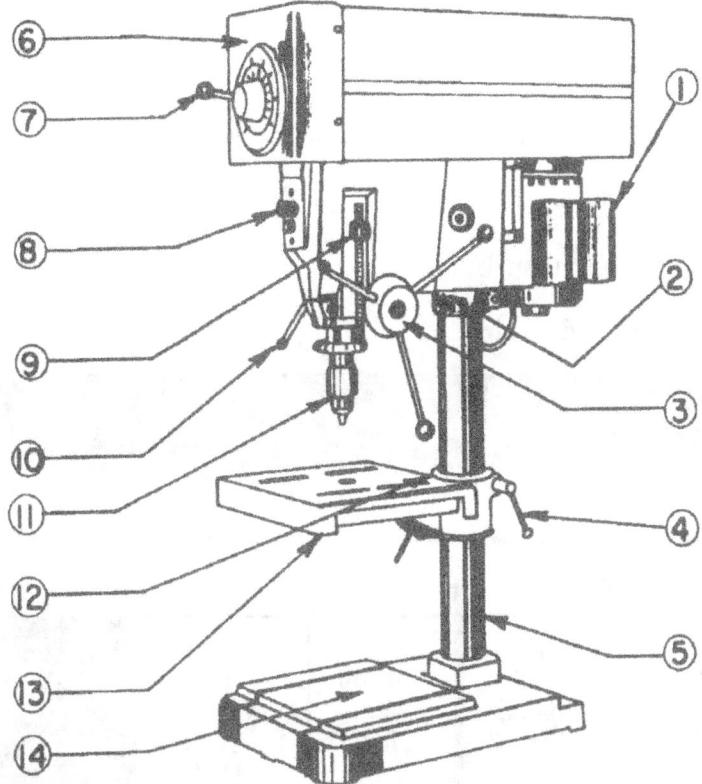

1. _____
2. _____
3. _____
4. _____
5. _____
6. _____
7. _____
8. _____
9. _____
10. _____
11. _____
12. _____
13. _____
14. _____

B. SAFE OPERATIONAL PROCEDURES:

1. General drilling:

    a. Use a drill gage to determine drill size.

    b. Be sure drill is sharp and ground properly.

    c. Be sure the material is clamped firmly.

    d. Feed drill into material at a constant rate.

    e. When drilling round stock, use a "V" block vise.

    f. Place work on a block of wood to prevent damage to the table.

2. Wood drilling:

    a. Using a combination square, draw two lines perpendicular to each other with their intersection being the center of the hole to be drilled.

b. Select proper drill speed.

   (1) Use 1250 RPM for most wood drilling up to 3/4" diameter.

   (2) Larger bits should be run at a slower speed.

   (3) Expansive bits will burn if run much over 600 RPM.

3. Metal drilling:

   a. Use a scribe or scratch awl to mark point for punch mark.

   b. Center punch a mark large enough to receive the point of the drill.

   c. Select the exact size of drill.

   d. Use a pilot hole for holes larger than 1/2".

   e. Select the proper drill sped for drill size.

DRILLING SPEEDS AND FEEDS							
Speed: Revolutions Per Minute							
	Cast Iron		Steel		Brass		
Diameter of Drill Inches	Carbon Steel Drill	High Speed Steel Drill	Carbon Steel Drill	High Speed Steel Drill	Carbon Steel Drill	High Speed Steel Drill	Feed Per Revolution Inches
	35 ft. per min.	70 ft. per min.	30 ft. per min.	60 ft. per min.	100 ft. per min.	200 ft. per min.	
1/4	535	1070	458	917	1528	3056	0.005
1/2	268	535	229	458	764	1528	0.008
3/4	178	357	153	306	509	1018	0.010
1	134	267	115	229	382	764	0.012

   f. Use cutting oil when drilling into steel.

   g. When drilling holes to be tapped, be sure to use the exact drill for the size tap selected. (See drill and tap charts). The hole should be approximately 1/16" smaller than the tap size.

C. **GENERAL SAFETY PRACTICES:**

   1. Wear eye protection at all times when operating the drill press.

   2. Do not operate the drill press without permission from the instructor.

   3. Clamp the material to be drilled securely to the table.

4. Place the long end of the piece being drilled to the left so it will hit the post and not the operator should the material slip and start rotating.

5. Be sure chuck is tight on drill and that the drill and chuck match.
   The chuck should always be tightened in all three tightening positions.

6. Always remove the chuck wrench from the chuck immediately after using it.

7. Be sure drill is sharp and operating at proper speed.

8. Never wear gloves while operating the drill press. They may get caught.

9. Do not wear loose-fitting clothes when operating the drill press.

10. Do not feed the drill faster than it can be easily cut.

11. Remove the chips from the machine with a brush.

12. Do not reach around the machine.

13. Keep long hair tied back or covered to keep it from getting wrapped around the chuck.

14. Do not talk to anyone while operating the machine.

15. Do not drill into a container that may have contained gasoline or other flammable materials.

16. Support long material being drilled.

17. Slow drill feed when it is breaking through the material to finish hole.

18. Hold round stock in a "V" block.

19. Clamp sheet metal between two blocks of wood and drill through wood and metal.

D. COMPLETION QUESTIONS:

1. A drill gage can be used to determine _____ _____.

2. A _____ block should be used to hold round stock.

3. A _____ _____ placed under the material being drilled will prevent damage to the table.

4. Expansive wood bits should not run faster than _____ RPM.

5. A _____ punch should be used before a hole is drilled in a piece of metal.

6. As a rule, larger drills run _____, while smaller drills run _____.

7. When drilling a hole in metal larger than 1/2", use a _____ _____.

8. _____ oil should be used when drilling steel.

9. When drilling a hole to be tapped, the drill size should be approximately 1/16" _____ than the tap size.

10. The long end of a piece of material should be kept to the _____ to prevent injury to the operator should it start to rotate.

PORTABLE DRILL

A. **PART IDENTIFICATION:**

Identify the circled parts on the portable drill illustrated below.

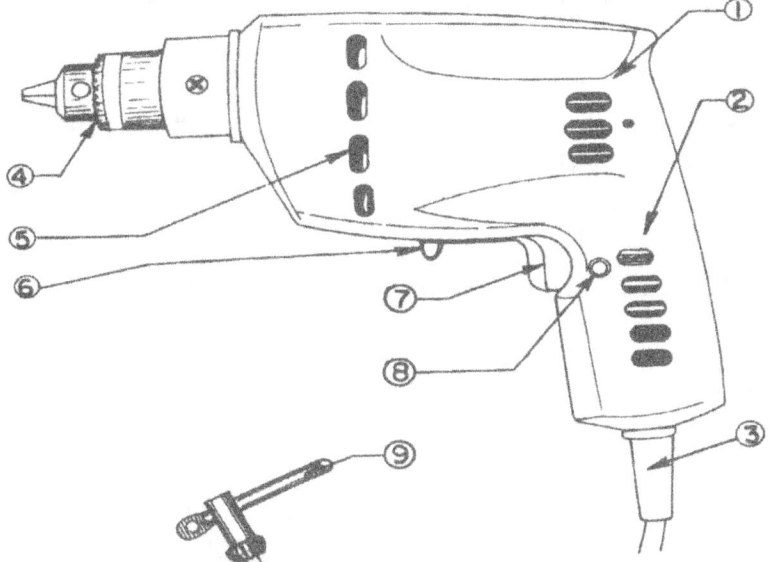

1. _____
2. _____
3. _____
4. _____
5. _____
6. _____
7. _____
8. _____
9. _____

B. **SAFE OPERATIONAL PROCEDURES:**

1. Select the proper twist drill for material to be drilled and make sure it is properly sharpened.

2. Tighten chuck securely in all three tightening locations so that the twist drill will run true.

3. Locate exact point of desired penetration and mark with a center punch or awl.

4. Securely fasten all materials so that they cannot twist or turn during the drilling process.

5. Small size twist drills should run at a faster RPM than large size twist drills.

6. Let drill come to full speed before starting to drill.

7. Drill with even, steady pressure and let the drill do the work.

8. When drilling deep holes, withdraw the drill several times to clear the cuttings or chips.

9. Hold drill at the correct angle while drilling.

10. As the twist drill starts to break through the material, ease off on the pressure to prevent grabbing or splintering.

11. Apply a suitable coolant or lubricant if necessary.

12. Clamp scrap stock on the back side of material to be drilled so it will not grab or splinter when the twist drill breaks through.

13. Raise drill soon after it starts cutting and inspect the impression to make sure it is centered.

14. Twist drills are made and sharpened to operate in the right hand direction.

15. Drill a pilot hole first when drilling large holes.

16. Disconnect the drill from the power source and remove the twist drill from the drill as soon as work has been completed.

17. Larger twist drills should be used at slower speeds.

18. Variable speed drills can operate from 0 to 1000 RPM.

C. **GENERAL SAFETY PRACTICES:**

1. Do not wear loose-fitting clothes while working with a drill.

2. Do not operate the portable drill without permission from the instructor.

3. Remove rings, ties, and watches. Roll up sleeves.

4. Wear a face shield or goggles while working with a drill.

5. Disconnect drill from power source before removing or installing twist drills.

6. Before connecting to power source, make sure the switch is in the "off" position.

7. Make sure the key has been removed from the chuck before starting the drill.

8. Be sure the drill is properly grounded when connected to the power outlet or is double insulated.

9. Do not use a drill which has a broken or damaged plug or cord.

10. While using a large drill, brace the body well to prevent injury.

11. Use a brush to remove chips.

12. Keep attention focused on the work.

13. Note position of cord to avoid drilling into it or getting it wrapped around the twist drill.

14. Make certain the drill will not injure someone working on the other side of the work.

15. Never operate the portable drill while standing on water or on a wet floor.

D. COMPLETION QUESTIONS:

1. Be sure the switch is _____ before connecting the drill to the power source.

2. A _____ or _____ can be used to establish the starting point for the twist drill.

3. Larger twist drills should be used at a _____ speed.

4. A pilot hole is used so less pressure is required when drilling _____ holes.

5. When drilling deep holes, withdraw the twist drill several times to clear the _____.

6. Disconnect the drill from _____ before removing or installing twist drills.

7. Check to see if the _____ has been removed from the chuck before starting the drill.

8. Use a _____ to remove chips.

9. Twist drills are made to operate in a _____ hand direction.

10. Variable speed drills can operate at speeds from _____ to 1000 RPM.

NOTES

NAILER AND STAPLER

A. PART IDENTIFICATION:

Identify the circled parts on the nailer and stapler illustrated below.

NAILER

1. _____
2. _____
3. _____
4. _____
5. _____
6. _____
7. _____
8. _____
9. _____
10. _____

STAPLER

1. _____
2. _____
3. _____
4. _____
5. _____
6. _____
7. _____
8. _____
9. _____
10. _____

B. SAFE OPERATIONAL PROCEDURES:

1. Study the operation, maintenance, and safety manual(s) for specific model and type of nailer or stapler.

2. Check the air pressure gage, lines, and all connections for leaks and proper operation.

3. Operating air pressure should be between 70 and 100 PSI, 90 PSI will produce best performance from pneumatic nailers and staplers. Never exceed 120 PSI for operating a nailer or stapler.

4. Check the air inlet plug (male connector) and air coupler (female connector) to make sure the plug and coupler match.

5. Connect only male connectors to the tool so high pressure air can be vented from tool when the line is disconnected. A female quick coupling could trap air in the tool leaving it live for one extra, unexpected shot.

6. Make all adjustments on the tool before loading nails or staples.

7. Lubricate the tool according to manufacturer's recommendation. If oil is added into air line with a lubricator, lubricating the tool will not be necessary; however, the lubricator should be checked on a daily use basis.

8. Select size and type of staples or nails for the machine and the job to be completed.

9. Follow correct procedures for loading the nailer or stapler. Be sure that air lines are disconnected for loading.

10. The nailer or stapler should never be used as a hammer or dropped; the tool housing may be cracked or weakened making it unsafe. Do not engrave or stamp the main housing, this could weaken the housing unit.

11. After the tool is cleaned, lubricated, adjusted, and loaded with the proper nails or staples, properly connect the tool to the air line for operation.

12. Carry the tool by the handle only, not by the air line, and keep fingers away from the trigger. Always keep nosepiece aimed toward the ground, and never pointed toward anyone.

13. The tool should be operated only when in contact with the workpiece. Use caution when nailing thin materials or near corners or edges to avoid driving the fastener through or away from the workpiece. The nailer or stapler should always be placed squarely on the surface or workpiece to be nailed to avoid the danger of fasteners ricocheting off the surface.

14. When not using the tool or if leaving the work area, disconnect it from the air line.

15. Disconnect the air line when attempting to clear a jam or when repairing the tool.

16. Do not attempt to adjust or remove the work contacting element. If it is not working correctly, do not use the tool. This is a "safety" on most nailers and staplers and it will not fire until the element contacts the workpiece.

17. When work is completed, disconnect the unit from the air line, shut off the air supply to the tool hose, and vent the compressed air from hose with an air nozzle.

18. Remove the fasteners from the tool, clean tool, and place in proper storage.

C. GENERAL SAFETY PRACTICES:

1. Wear industrial quality eye protection, hearing protection, and proper clothing when working with these tools.

2. Obtain permission from the instructor to use the air nailer or stapler.

3. Always assume the tool is loaded.

4. Never point the tool at anyone, even if the air line is disconnected.

5. Disconnect air line when not in use or when leaving the work area.

6. Carry the tool only by the handle, not by the hose.

7. Make sure the work contacting element is in good working order and in contact with the workpiece before depressing the trigger.

8. Use matching air connectors and couplings.

9. Select correct nails or staples manufactured for the tool.

10. Do not remove or tamper with the work contacting element.

11. Use only regulated air pressure; never use bottled air or gases to power the tool.

12. Use only recommended air pressure; under pressure operation may be as dangerous as over pressure operation.

13. The nailer or stapler is a tool, not a toy.

D. COMPLETION QUESTIONS:

1. The recommended air pressure for most air nailers and staplers is _____ PSI.

2. Always assume the tool is _____.

3. The power source for the nailer or stapler should be _____ _____, not bottled air or gas.

4. Carry the tool only by the _____ with the _____ in a safe position.

5. The _____ _____ should always be in contact with the workpiece before the tool is fired.

6. Only _____ connectors should be fitted to the tool.

7. Stamping or engraving on the _____ _____ could weaken the tool causing it to be unsafe.

8. The fasteners are held in the _____ portion of the tool.

9. The tool should be _____ daily unless oil is added directly into the air line.

10. Another name for the air tool is a _____ tool.

NOTES

GRINDER

A. **PART IDENTIFICATION:**

Identify the circled parts on the grinder illustrated below.

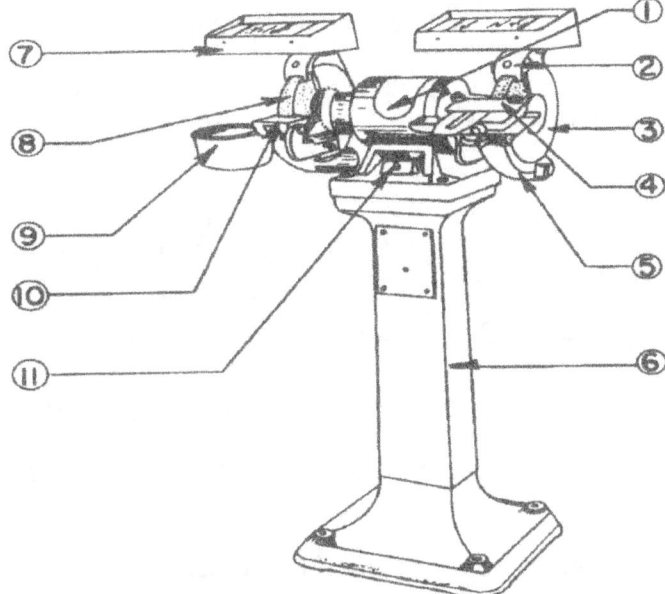

1. _____
2. _____
3. _____
4. _____
5. _____
6. _____
7. _____
8. _____
9. _____
10. _____
11. _____

B. **SAFE OPERATIONAL PROCEDURES:**

1. Tool sharpening:

    a. Selecting the grinding wheel:

       (1) Select a 60-grit wheel for general tool sharpening.

       (2) A soft grinding wheel is recommended for grinding hard materials.

       (3) The grinding wheel should fit inside the housing.

       (4) The arbor hole size should be .002 inches larger than the shaft size.

       (5) The grinder speed should not exceed the speed listed on the side of the grinding wheel.

    b. Maintaining the grinding wheel:

       (1) Use a wheel dresser to remove the glaze on the wheel after grinding soft material.

       (2) Use a wheel dresser for straightening a grooved, rounded, or out-of-round wheel.

(3) Before dressing the wheel, adjust the tool rest on the grinder to a point even with the center of the wheel.

(4) Start the grinder. Place the wheel dresser on the tool rest, gripping it with both hands.

(5) Move the wheel dresser back and forth across the wheel. Do not let the dresser pass off the edge of the wheel.

(6) Remove just enough material to clean and straighten the wheel.

(7) Use a combination square to make sure the face of the wheel is square with the side.

c. Sharpening a tool:

(1) Adjust the tool rest so that it is not more than 1/8 inch from the wheel and slightly above the center.

(2) Stand to one side and start the grinder.

(3) Allow the machine to come up to speed before starting to grind.

(4) Carefully place the tool against the grinding wheel so its cutting edge is against the direction of rotation.

(5) Move the tool slowly across the face of the wheel.

(6) Do not allow the tool to overheat. Dip it in water or use a very light feed and stop to allow it to cool in the air.

C. GENERAL SAFETY PRACTICES:

1. Wear eye protection and proper clothing at all times when operating a grinder.

2. Do not operate the grinder without permission from the instructor.

3. Be sure the housing is in place around the grinding wheel. It is there for protection in case the grinding wheel should break.

4. Do not use a wheel that vibrates excessively.

5. Never use a grinding wheel that is cracked or broken.

6. Do not use the grinder if the light is not adequate.

7. Wear a face guard even though the grinder has a glass shield.

8. Do not hold material being ground in such a way that fingers may contact the wheel surface.

9. Avoid using the side of the wheel for rough grinding. This may place too much stress on the wheel and cause it to break.

10. Do not force material into the wheel.

11. If the wheel cuts slowly or vibrates because of uneven wear, use a wheel dresser to expose a new cutting surface.

12. Do not operate the grinder if another person is standing nearby.

13. Never remove the paper from the sides of the grinding wheel.

14. Do not use a grinding wheel that is worn to less than 1/2 its original diameter. For example, an 8" wheel should not be used at less than 4" in diameter.

15. Do not leave the grinder without turning it off.

D. COMPLETION QUESTIONS:

1. Soft stones are recommended for grinding _____ materials.

2. The arbor hole size should be _____ inch (es) larger than the shaft size.

3. A _____ _____ is used to straighten a grooved, rounded, or out-of-round grinding wheel.

4. The wheel dresser should not pass off the _____ of the grinding wheel.

5. A _____ _____ can be used to determine if the face of the wheel is square with its side.

6. Before sharpening a tool, adjust the tool rest so it is not more than _____ inch(es) from the wheel.

7. The _____ _____ of the tool being sharpened should be against the direction of rotation.

8. The tool should be moved slowly across the _____ of the wheel.

9. The tool can be dipped in _____ to keep it from overheating.

10. Stand _____ _____ _____ when starting the grinder.

NOTES

BELT GRINDER

A. **PART IDENTIFICATION:**

Identify the circled parts on the belt grinder illustrated below.

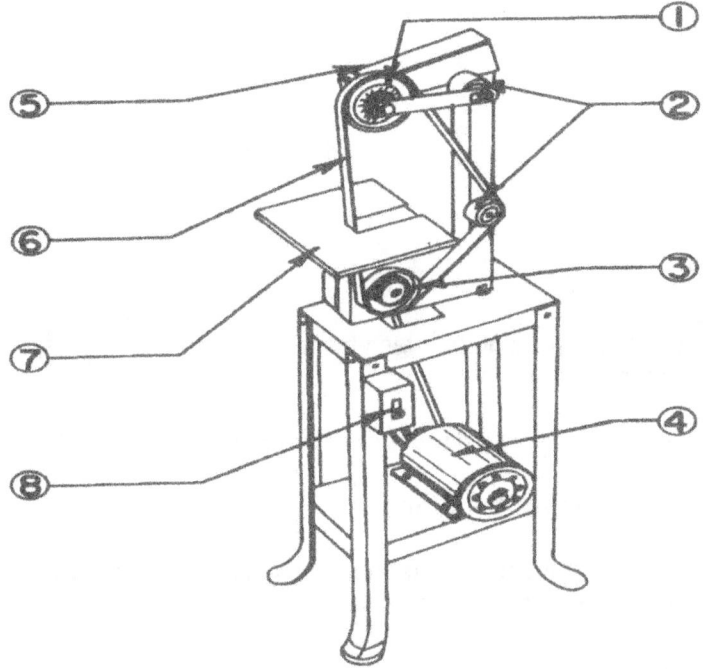

1. _____
2. _____
3. _____
4. _____
5. _____
6. _____
7. _____
8. _____
9. _____
10. _____

B. **SAFE OPERATIONAL PROCEDURES:**

1. Check the belt to see if it is torn or otherwise damaged.

2. Make sure the belt runs on the center of the pulleys.

3. Select a new belt if needed and mount it on the pulleys.

4. Check the belt direction before attempting to grind. The belt is spliced for operation in only one direction. If the belt is mounted wrong, it will tear.

5. Adjust the table so it clears the belt by 1/8".

6. Check the belt for tightness. If the belt is too loose, it will pile up in front of the work. If the belt is too tight, there will be unnecessary wear on the bearings.

7. If the belt does not run centered on the pulleys, the idle pulley will need to be tilted to make one side of the belt tighter than the other. The belt will run toward the looser side of the pulleys.

8. Start the motor before applying the work to the belt.

9. Feed metal slowly; do not force it.

10. Keep the work square on the belt.

11. Hold the work so it will not run off the edge of the belt or get caught under the belt.

12. Use the entire width of the belt.

13. Hold sharp pointed metal so it will not jam and catch into the belt.

14. Turn off the grinder before leaving it.

15. Keep a pail of water nearby in which to cool metal.

16. When grinding long material, support one end on a solid rest or have someone help by holding it.

C. GENERAL SAFETY PRACTICES:

1. Wear safety glasses and a face shield to protect eyes and face from flying chips when operating a belt grinder.

2. Never operate the belt grinder without permission from the instructor.

3. Do not talk to anyone while operating the belt grinder.

4. Be sure the main power switch is off or the cord is disconnected before making any adjustments or repairs on the grinder.

5. Keep fingers away from the belt.

6. Do not wear loose fitting clothing. It may get caught in the belt.

7. Use vise grips to hold small pieces.

8. Do not leave the grinder until it has been shut it off and has stopped running.

9. Clean off chips with a brush.

10. Do not rub eyes before washing any chips and grit off hands. Small pieces of metal may get into eyes and cause serious injury.

11. Do not use the grinder if the light is not adequate.

12. Keep the floor around the grinder clear of scraps and other material.

13. Keep flammable materials, such as gasoline, away from the belt grinder.

D. COMPLETION QUESTIONS:

1. The belt should run on the _____ of the pulleys.

2. The _____ is spliced for operation in only one direction.

3. The table should clear the belt by _____ inches.

4. If the belt is too loose, it will _____ _____ in front of the work.

5. The belt will run toward the _____ side of the pulleys.

6. If the belt is too tight, the _____ will wear excessively.

7. _____ the motor before applying work to the belt.

8. The _____ pulley can be moved to adjust the tightness of the belt.

9. If the belt runs in the wrong direction, it will _____.

10. The work should not _____ _____ the edge of the belt.

NOTES

PORTABLE GRINDER

A. **PART IDENTIFICATION:**

Identify the circled parts on the portable grinder illustrated below.

1. _____
2. _____
3. _____
4. _____
5. _____
6. _____
7. _____
8. _____
9. _____
10. _____

B. **SAFE OPERATIONAL PROCEDURES:**

1. Select a grinding wheel that will fit the arbor shaft.

2. Be sure to have a paper washer on both sides of the grinding wheel.

3. Always check the RPM range of the wheel as compared to the speed of the grinder to protect the wheel from exploding if too high an RPM is used.

4. Check the wheel for cracks by holding it up with the thumb and forefinger and striking it lightly with a hard object. The lack of a ringing sound indicates a cracked grinding wheel.

5. Mount the wheel on the arbor. Be sure the wheel fits the arbor shaft.

6. Be sure the concave side of a large washer is in contact with both sides of the wheel.

7. Firmly tighten the nut.

8. Hold the grinder with both hands.

9. Be sure the wheel guard is in place.

10. Be sure no one is in line with the wheel before starting the motor.

11. Turn the grinder motor on and off and check to see if the wheel vibrates excessively or does not run round and true.

12. Be sure the grinding wheel is at room temperature. A very cold wheel may break when used.

13. Run the grinder at full speed for one minute after a wheel has been mounted or has been roughly treated. If a wheel is going to break, it will usually do so when the grinder is first turned on.

14. Be sure the work is held firmly.

15. Feed the grinding wheel lightly into the work after the motor has come up to full speed.

16. Do not force the wheel into the work and cause the wheel speed to be reduced.

17. Do not lay the grinder down until the wheel has stopped turning.

18. Lay the grinder on its rest plate so nothing touches the grinding wheel while it is not in use.

19. Return the portable grinder to its case or cabinet after use.

C. **GENERAL SAFETY PRACTICES:**

1. Always wear a plastic face shield as well as safety glasses when operating a portable grinder.

2. Do not use the portable grinder without permission from the instructor.

3. Be sure the switch is off and the cord is disconnected from the power source before making any adjustments, lubricating, inspecting, or changing grinding wheels.

4. Never use a cracked grinding wheel or one that vibrates excessively.

5. Be sure the grinder is properly grounded or double insulated.

6. Use only wheels that are designed to operate at the speed indicated on the grinder nameplate.

7. Be sure the wheel guard is in place.

8. Do not direct the discharge at anyone as the sparks can cause burns and small pieces can become embedded in the skin.

9. When grinding small pieces, be sure they are held securely in a vise or clamped to the table.

10. Do not talk to anyone while operating the grinder.

11. Do not use the grinder in areas where flammable materials are kept.

12. Do not wear loose-fitting or frayed clothing.

13. Make sure the floor around the work area is clean and free of other materials before operating the grinder.

14. Keep the power cord away from the grinding wheel.

15. Always hold the grinder with both hands.

16. Always make sure the switch is off when the grinder is not in use so the grinder will not start when the cord is connected to a power source.

17. Be sure the grinder does not exceed the speed stamped on the side of the grinding wheel.

18. Never operate the portable grinder while standing in water or on a wet floor.

19. Never use a grinding wheel when it is less than 1/2 of its original diameter.

D. COMPLETION QUESTIONS:

1. The grinder speed should never _____ the speed indicated on the side of the grinding wheel.

2. A cracked grinding wheel will not have a _____ sound when struck lightly with a hard object.

3. The grinder should be lying on its _____ when not is use.

4. Do not lay the grinder down until the _____ has stopped turning.

5. Hold the grinder with _____ _____.

6. Be sure the _____ guard is in place.

7. Run the grinder at full speed for _____ minute(s) after mounting a grinding wheel.

8. Do not _____ the wheel into the work.

9. A very _____ wheel may break when first turned on or used.

10. The _____ side of the large washers should be next to the grinding wheel.

NOTES

METAL CUTTING BAND SAW

A. **PART IDENTIFICATION:**

Identify the circled parts on the metal cutting band saw illustrated below.

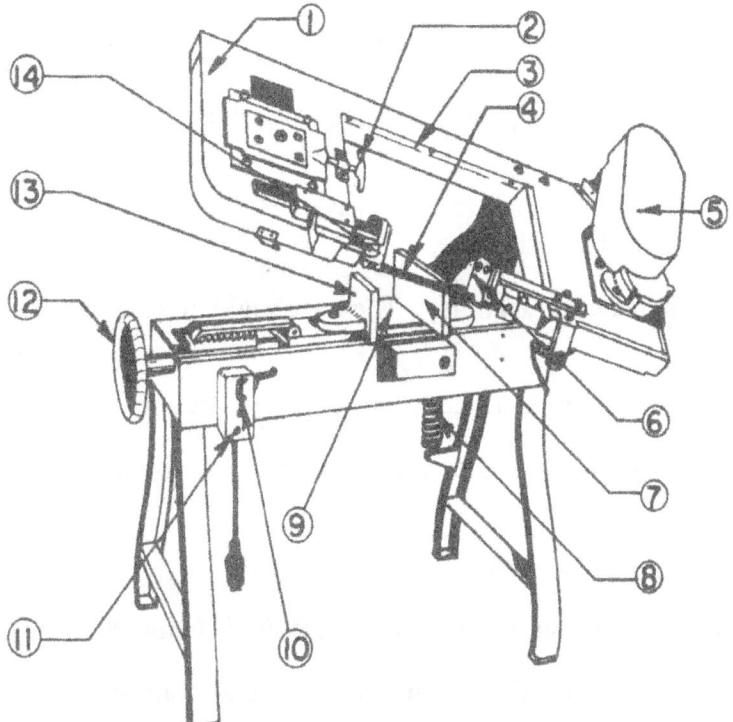

1. _____
2. _____
3. _____
4. _____
5. _____
6. _____
7. _____
8. _____
9. _____
10. _____
11. _____
12. _____
13. _____
14. _____

B. **SAFE OPERATIONAL PROCEDURES:**

1. Replacing blade:

    a. Select a blade that is sharp and in good condition with 14-18 teeth per inch for thin wall tubing.

    b. Place the blade in the saw with the exposed teeth pointing toward the motor.

    c. Tighten the blade until it is snug. Do not over-tighten.

    d. When a worn or broken blade is replaced, the stock being cut should be turned over. A new blade is thicker and will be damaged if allowed to enter an old saw kerf.

2. Placing stock in the vise of a metal cutting band saw:

    a. Clamp all stock firmly on the vise to prevent breaking the blade.

    b. Clamp angle iron in the vise with the legs down.

    c. Clamp rectangular material with the widest side toward the blade.

    d. Support long stock.

    e. When cutting short material, a block of equal width can be placed in the opposite end of the vise jaws. This will allow the vise to grip the stock tighter.

    f. Never use a piece of pipe or a wrench as a lever to help tighten the vise.

    g. Use a cut-off gage when cutting short pieces of stock the same length.

3. Sawing:

    a. Properly clamp the material in the vise.

    b. Lower the blade and check for accuracy. The cut should be made in the waste stock.

    c. Move the adjustable blade guide as close to the work as possible.

    d. Raise the blade, start the saw, and then lower the blade gently onto the work.

    e. Do not force the saw into the work.

    f. Do not attempt to break off the material before the cut is complete.

    g. Wait for the machine to automatically shut off at the end of the cut.

    h. Release the metal from the vise.

    i. Clean the area before leaving. Remove all scrap from the floor.

C. GENERAL SAFETY PRACTICES:

1. Do not operate the metal cutting band saw without permission from the instructor.

2. Wear safety glasses and keep hands away from eyes. Small pieces of metal rubbed into the eyes can cause pain and serious injury to the eyes.

3. Keep hands a reasonable distance from the blade when cutting.

4. Keep the adjustable blade guide as close as possible to the material.

5. Be sure the metal to be sawed is held securely in the vise.

6. Make sure the pulley shield is always in place before use.

7. Do not wear loose-fitting clothes. Keep sleeves rolled up above the elbows.

8. If the blade breaks, do not attempt to stop the machine; it will stop automatically.

9. Shut off main power switch before making adjustments or repairs on the machine.

10. Allow the blade to feed in the work slowly when starting a cut.

11. Do not force the speed of cutting by applying pressure.

12. Support long material.

13. Do not leave saw running unattended.

14. Remove the burr left by the saw by filing or grinding.

15. After the saw has stopped, use a brush, not bare hands, to remove chips from the machine.

16. Remove scrap material from the floor.

D. **COMPLETION QUESTIONS:**

1. When replacing the blade in a metal cutting band saw, the teeth should be pointing toward the _____.

2. When a worn or broken blade is replaced, the stock being cut should be _____ _____.

3. All stock should be firmly clamped in the _____ to prevent breaking the blade.

4. Angle iron should be clamped in the vise with the legs _____.

5. The widest side of rectangular material should be toward the _____.

6. A _____ _____ gage can be used when cutting off several pieces of stock the same length.

7. The cut should be made in the _____ stock.

8. Do not _____ the saw into the work.

9. Do not break off the _____ before the cut is complete.

10. Before starting the saw, the adjustable blade _____ should be moved as close as possible to the work.

NOTES

PORTABLE METAL CUTTING BAND SAW

A. PART IDENTIFICATION:

Identify the circled parts on the portable metal cutting band saw illustrated below.

1. _____
2. _____
3. _____
4. _____
5. _____
6. _____
7. _____
8. _____
9. _____
10. _____
11. _____
12. _____

B. SAFE OPERATIONAL PROCEDURES:

1. Study the operation, maintenance, and safety manual(s) for the specific model and type of saw.

2. Select a blade that is sharp and has the correct teeth/inch based on the job. For thicker materials, 1" or more, select a blade having 6-8 teeth per inch, for 1/4" to 3/4" materials use a blade with 10-18 teeth per inch, and for less than 1/4" thickness use a blade having 24 teeth per inch. Select a blade which will allow at least two teeth to be engaged in the material thickness. The thinner or the harder the material, the finer the blade teeth. The thicker or the softer the material, the coarser the blade teeth.

3. To remove the worn blade, turn the blade release handle clockwise to release the blade tension. Remove the blade first from the pulleys and then from the blade guides.

4. Before installing a blade, clean chips and wax from the blade guides and pulley tires. Insert blade in blade guides first and then position on pulleys. Place the blade in the saw with the exposed teeth pointing toward the handle or back part of the saw.

5. When the new blade is in place, turn the blade release handle counter-clockwise. Check the blade to see that it is snug and fitting into the guide rollers.

6. When a worn or broken blade is replaced, do not start a new blade in a partially cut workpiece as the set of a new blade is thicker and will be damaged if allowed to enter an old saw kerf.

7. If the saw has two-speed or variable speed control, select the correct saw speed according to the type of material to be cut. Blade speed is SFPM (Surface Feet Per Minute) and will range from 80 to 240 SFPM. The general rule is the harder the stock such as high speed steel, stainless steel, chrome, or tungsten steel, the slower the blade speed (80-150 SFPM). Metals such as aluminum, brass, copper, soft bronze, and low carbon steel may be cut at high speed, 150 to 240 SFPM.

8. If the saw has an adjustable blade guide, adjust blade guide according to size of stock, keeping guide as close to stock as possible.

9. Never use liquid coolant on the band saw as coolant could damage the blade guide bearings or rubber tires on the pulleys. Blade wax may be used when cutting aluminum, brass, or thicker materials. Cast iron and steel should be cut dry.

10. Clamp all stock firmly in a vise or by other clamping methods to prevent breaking blades or injury to the operator. Never attempt to saw material being held by someone.

11. Cut angle iron with legs down and rectangular stock with widest side toward the blade.

12. To prevent pinching or twisting the saw blade or dropping the waste stock on someone, support the waste stock with a roller stand, or have a helper hold it.

13. Mark the stock to be cut and always saw on the waste side of the line.

14. Check the electrical cord to see that it is in good condition and use only electrically grounded outlets.

15. As the operator, assume a position of good balance and comfort for holding and controlling the saw.

16. Grip the saw by the front and rear handles making sure the trigger switch remains in the "off" position until ready to begin the saw cut.

17. Position the saw so the saw work stop is against the stock being cut. If not, the saw will quickly pull itself back against the stop as it is started due to the direction of blade rotation. This quick movement could damage the blade or jerk the saw from the operator's hands.

18. Hold the saw so the blade is just above the stock before depressing the trigger switch. Depress the switch and lower the blade gently onto the workpiece.

19. Do not force the saw into the workpiece. Hold the saw squarely and firmly against the workpiece allowing the saw to do its job at its own pace with only the weight of the saw exerting pressure down on the workpiece. Additional pressure will slow down the speed of the blade and reduce cutting efficiency. Do not rock the saw in an attempt to get it to saw faster as this will shorten blade life.

20. Do not attempt to break off the material before the cut is compete.

21. Hold firmly onto the saw as the cut is nearing completion to keep it from dropping or hitting the workpiece as the cut is completed.

22. Release the trigger switch and allow the saw to stop before setting down on the bench or other safe resting area. Disconnect the saw from the electrical outlet.

23. Release the metal from the vise, clean the area, and remove all scrap from the floor.

24. Clean the saw and return it to its proper storage when the job is completed.

C. GENERAL SAFETY PRACTICES:

1. Obtain permission from the instructor to operate the portable metal cutting band saw.

2. Wear industrial quality eye protection and proper clothing. Keep hands away from eyes as small pieces of metal filings may be rubbed into the eyes causing pain or injury.

3. The operator should always have both hands on the saw and make sure helpers are at a safe distance from the saw blade.

4. Keep the adjustable blade guide as close as possible to the stock being cut.

5. Keep blade guards in place at all times.

6. Secure metal stock to be cut in a vise or other clamping methods. Never attempt to saw stock being held by a helper.

7. If a blade breaks, do not attempt to stop the saw. Release the trigger switch and it will stop on its own.

8. The saw should be disconnected from the electrical power source before changing blades or making any adjustments.

9. Make sure the saw is properly grounded and that it is connected to a grounded electrical source having sufficient amperage for the saw.

10. Start the saw before the blade touches the stock to be cut. Make sure the work stop is touching the workpiece.

11. Do not force the speed of cutting by applying pressure to the saw other than the weight of the saw.

12. Support long material. Do not saw between two supports causing the material to bind or pinch the saw blade.

13. Remove burrs left by the saw on the stock by filing or grinding.

14. Use a brush, not bare hands, to remove iron filings from the saw or work area.

15. When the job is completed, remove all scrap materials from the floor.

D. COMPLETION QUESTIONS:

1. When replacing the blade in the portable metal cutting band saw, the teeth should be pointing toward the _____.

2. When cutting stock that is 1/2" thick, a blade having _____ teeth per inch should be selected.

3. The general rule for blade speed is the harder the stock being cut, the _____ the blade speed, so when cutting carbon steel, the blade speed should be _____ to _____ SFPM.

4. A new blade could be damaged if allowed to enter a partially sawed workpiece because the saw _____ is _____ than the new blade will cut.

5. If cutting high speed steel, select a blade speed of _____ SFPM.

6. Angle iron should be cut with the angle legs _____.

7. The _____ _____ handle is used to release pressure on the blade to replace a worn blade.

8. Mark the stock to be cut and always saw on the _____ side.

9. The _____ _____ should be held firmly against the stock before depressing trigger switch to avoid the saw from being jerked from the operator's hands.

10. The saw should be _____ before the blade is allowed to touch the workpiece.

METAL LATHE

A. PART IDENTIFICATION:

Identify the circled parts on the metal lathe illustrated below.

1. _____
2. _____
3. _____
4. _____
5. _____
6. _____
7. _____
8. _____
9. _____
10. _____
11. _____
12. _____
13. _____
14. _____
15. _____
16. _____
17. _____
18. _____
19. _____
20. _____

B. SAFE OPERATIONAL PROCEDURES:

1. Facing off round stock:

    a. Select a piece of cold rolled steel 1" diameter by 3"-12" in length.

    b. Mount stock in lathe using a three-jaw universal chuck with about 1-1/8" of stock sticking out of the chuck.

    c. Select a correctly sharpened right-handed general purpose turning tool.

d. Mount a left-handed tool holder in the tool post, insert the turning tool in the tool holder, and tighten the set screws with the cutting edge at the same level as the center of the stock.

e. Rotate spindle by hand to insure its free movement.

f. Set spindle speed at 600 RPM.

g. Start machine.

h. Face off end of stock by feeding the tool into the center of the stock with the carriage handwheel and then slowly moving the tool back toward the operator using the cross slide handwheel.

2. Drilling center holes with a lathe chuck:

   a. Select a universal three-jaw chuck. Protect ways with a board.

   b. Wipe the cone surfaces clean with a rag and mount the chuck on the headstock.

   c. Securely fasten a short shaft in the chuck with 1 to 2 inches extending to the right of the jaws.

   d. Face off the end of the shaft as described above.

   e. Mount the drill chuck in the tailstock after wiping the tapered surfaces clean.

   f. Securely mount a combination center drill and countersink in the drill chuck. Tighten chuck in all three tightening positions with proper chuck wrench.

   g. Move the combination center drill and countersink close to the faced off stock by sliding the tailstock. Clamp down the tailstock.

   h. Set the spindle speed at 150 RPM.

   i. Start the motor.

   j. Use the tailstock handwheel to slowly feed the combination center drill and countersink into the stock.

   k. Stop drilling when the outside diameter of the countersink hole is 3/16 inch if the diameter of the countersink body is 1/4 inch.

   l. After the center holes have been properly drilled in both ends of the stock, it is ready to be mounted between centers.

3. Mounting the work between centers:

   a. Lubricate center holes with a high pressure lubricant.

   b. After wiping clean the tapered surfaces, wedge a dead center in the headstock, and a live or dead center in the tailstock.

c. Mount a face plate on the headstock.

d. Slide a lathe dog over the stock.

e. Hold the stock between the centers with one hand and move the tailstock up near the end of the stock.

f. Clamp the tailstock to the bed.

g. Snug the centers into the center holes.

h. Place the lathe dog into a notch in the face plate.

i. Tighten the set screw in the lathe dog.

j. The lathe dog should move freely back and forth in the face plate notch.

k. While moving the dog back and forth, tighten the centers by rotating the tailstock handwheel until a slight tension is felt.

l. Clamp tailstock spindle.

C. **GENERAL SAFETY PRACTICES:**

1. Do not operate the metal lathe without permission from the instructor.

2. Wear eye protection and remove loose clothing when operating the metal lathe.

3. Roll shirt sleeves above the elbows before operating the metal lathe.

4. Remove all jewelry including rings, wristwatches, etc.

5. Remove all wrenches, oil cans, and other materials from the work area before starting the machine.

6. Be sure the chuck is tightly mounted on the spindle.

7. Be sure the stock is tightly mounted in the chuck.

8. Do not leave a chuck wrench in the chuck at any time.

9. The operator should always start and stop the machine.

10. Never reach across the work while the machine is running.

11. Use a brush or hook to remove all chips.

12. Stop the machine before making any adjustments.

13. Stop the machine for all measurements.

14. Stop the feed before the tool reaches the jaws of the chuck.

D. **COMPLETION QUESTIONS:**

1. A _____ _____ tool holder works best for facing up round stock.

2. When facing round stock, it should extend _____ inches to the right of the chuck jaws.

3. Before starting the machine, the chuck should be _____ by hand to insure its free movement.

4. A _____ hand general purpose turning tool should be used to face off round stock.

5. The spindle should rotate at about _____ RPM when facing off round stock.

6. The _____ _____ handwheel and the _____ handwheel should be used to feed the turning tool into the work.

7. The _____ post holds the tool holder in place.

8. The _____ surfaces should be wiped clean before the drill chuck is mounted.

9. The spindle speed should be about _____ RPM for drilling center holes.

10. Dead centers should be lubricated with a mixture of _____ _____ and _____.

AIR IMPACT WRENCH

A. PART IDENTIFICATION:

Identify the circled parts on the air impact wrench illustrated below.

1. _____
2. _____
3. _____
4. _____
5. _____
6. _____
7. _____
8. _____
9. _____
10. _____
11. _____
12. _____

B. SAFE OPERATIONAL PROCEDURES:

1. Study the operator's manual that accompanies the specific model and make of air impact wrench.

2. The air impact wrench is powered by regulated air and is designed to be used to loosen and tighten nuts, bolts, and screws to a predesigned torque value.

3. Before opening the main air valve, check all hoses, pipe fittings, pressure gauges, and regulators for leaks or damage.

4. Operating air pressure should be no lower than 70 PSI or higher than 120 PSI with 90 PSI being the recommended air line pressure.

5. Check the CFM (cubic feet/minute) volume of the air supply. Most impact wrenches require 3 to 5 CFM for proper operation.

6. Check the air inlet plug (male connector) and air coupler (female connector) to make sure the plug and coupler match.

7. Select correct drive socket for job to be completed. Do not use regular sockets as they are not designed for impact wrench application. Special thick wall sockets are manufactured for impact wrench use.

8. Make all adjustments on wrench with the air line disconnected. Adjust the output torque control for desired bolt torque. Adjust the reversing valve for tightening or loosening fasteners.

9. Lubricate the impact tool according to manufacturer's recommendation. If an air line oil lubricator is used, check for oil supply and proper operation. If oil is added into air line, direct lubrication of the tool is not necessary.

10. When the impact wrench is properly adjusted and lubricated, connect to the air line.

11. Air impact wrenches should not be operated, other than a quick burst without contact to the fastener. Operation without contact to fastener could cause the socket to fly from the drive.

12. Make sure the workpiece containing fasteners to be tightened or loosened is secured in a vise or held by appropriate means, never by hand or by a helper.

13. When tightening or loosening fasteners, use the tool in short bursts to avoid stripping or as a check for the correct rotation.

14. If more or less tightening torque is required, adjust the output torque control screw accordingly.

15. When work is completed, shut off the air supply to the tool hose and vent compressed air in hose by squeezing the trigger.

16. Disconnect the air impact wrench from the air supply line, remove socket, clean the tool, and place in proper storage.

C. GENERAL SAFETY PRACTICES:

1. Wear industrial quality eye protection, hearing protection, and proper clothing when operating the air impact wrench.

2. Obtain permission from the instructor to use the air impact wrench.

3. Make all adjustments, lubricate, and clean the tool before connecting to the air supply line.

4. Disconnect the wrench from the air supply line when not using or when leaving the work area.

5. Carry the tool only by the handle and not by the air supply hose.

6. Select impact type sockets designed for the type of work to be completed.

7. Make sure air connectors (plug and coupling) are matched.

8. Use only regulated air; never use bottled air or gases to power the impact wrench.

9. Use only recommended air pressure; under-pressure operation may be as dangerous as over-pressure operation.

10. Make sure the workpiece is secured before attempting to loosen or tighten fasteners.

11. Operate the air impact wrench only in short bursts and only when making contact with fasteners.

12. Shut off the air line and bleed the air supply by squeezing the trigger before disconnecting the tool from the air supply line.

D. **COMPLETION QUESTIONS:**

1. The recommended air pressure for operating the air impact wrench is _____ PSI.

2. The air supply volume in CFM for most air impact wrenches is _____ to _____ CFM.

3. The air impact wrench should be carried by the _____ and not by the _____.

4. The _____ valve on the wrench is adjusted to change rotation.

5. The _____ _____ control is adjusted to obtain the correct tightening of fasteners.

6. The air impact wrench should be _____ daily unless oil is added directly into the air supply line.

7. The air impact wrench should only be operated with _____ air and not on bottled air or gases.

8. Specially designed impact _____ should be used as regular sockets have _____ walls which could be damaged during use and cause injury.

9. The air wrench should only be operated for short _____ and only when in _____ with fastener.

10. Only the _____ plug connector should be connected to the tool while the _____ coupler should be connected to the air supply line.

NOTES

GAS FORGE

A. PART IDENTIFICATION:

Identify the circled part on the gas forge illustrated below.

1. _____
2. _____
3. _____
4. _____
5. _____
6. _____
7. _____
8. _____
9. _____
10. _____
11. _____
12. _____

B. SAFE OPERATIONAL PROCEDURES:

1. Ignition instructions for spark ignited forge (follow manufacturer's instructions):

    a. If the forge has been in operation, always wait at least five (5) minutes between shutdown and starting up of the forge.

    b. Set air control halfway between open and closed positions. Set gas control to closed position.

    c. Using the lid handle, swing the lid toward the back side of the furnace so that it is not over the top slot.

    d. Depress and release the start button. The blower motor will start running, the red light will come on, and if the room is not too noisy, the spark plug igniter "buzzing" will be heard. The igniter will stay on for about 1-1/2 minutes, so the starting cycle must be completed during this time.

    e. Depress and <u>hold in</u> the igniter button. Slowly, turn the gas control toward the open position until the burners ignite. Then turn the control slightly past this position to obtain a steady roar from the burners. After about twenty (20) seconds, the red light will go out and the ignition button can be released. If the lighting cycle was not completed in 1-1/2 minutes, a thermal relay will shut off the gas and the spark igniter. Push the "stop" button, wait five (5) minutes, then repeat steps b, c, d, and e to start the forge.

  f. After the forge has been started, adjust the gas control to give a sharp tail of flame that extends just above the top of the forge. Work can be placed in the flame. The work rack at the front of the furnace can be slid out to support the work. The lid should be centered over the top slot, using the lid handle.

  g. To increase the amount of gas, turn the gas control toward the open position to get a higher flame. Then turn the air control handle toward the open position to obtain the sharp tail of flame. Repeat these steps until the desired or maximum gas input is reached.

  h. To decrease the amount of gas, turn the gas control handle toward the closed position until the sharp tail of flame almost disappears. Then turn the air control toward the closed position until the tail of flame reappears. Repeat this procedure until the desired or minimum gas input is reached.

  i. If the gas is turned too high causing a high, lazy flame or too low causing an intermittent flame, the thermocouple may cool down and shut off the burners. Push the stop button, wait five (5) minutes, and restart the forge.

2. Shutting down the spark ignited forge:

  a. To shut down the forge, turn gas control to the closed position and push the stop button.

  b. Do not attempt to speed the cooling process in the fire box by restarting the blower. The firebrick should cool slowly and evenly to prevent cracking.

3. Ignition instructions for the manually ignited forge:

  a. Pivot lid back until it no longer covers top slot.

  b. Turn solenoid gas control switch to "off" position.

  c. Plug in blower motor. Blower motor should start running.

  d. Turn valve in gas line to 1/2 open position.

  e. Ignite a fuel-soaked wick held in tongs. Hold burning wick in fire box.

  f. Keeping hands, face, and body away from top slot, turn solenoid gas control switch to "on" position. When forge has ignited, place remaining wick in a metal container and cover with a metal lid to extinguish flame.

  g. Adjust flame height to top of the forge with the gas line valve. Minor air adjustments may be made by rotating the disc which is located in front of the blower air inlet.

4. Shutting down the manually ignited forge:

  a. Turn solenoid gas control switch to "off."

  b. Close the gas line valve.

  c. Unplug the cord to the blower motor. Failure to stop blower may result in unequal cooling of firebrick and cracking.

C. **GENERAL SAFETY PRACTICES:**

1. Do not operate the gas forge without permission from the instructor.

2. Keep area near forge cleared of everything except tong rack and bucket of cool water to prevent accidental tripping of individuals using the forge.

3. All flammable objects and materials must be kept at a safe distance from the hot metal working area.

4. Adjust lid to proper height prior to igniting forge.

5. Wear safety shield over face to protect face and eyes from flame, sparks, and heat.

6. Wear heavy leather gloves when working with hot metals. Exercise great care in handling metal even with gloves.

7. Never wear oily or loose fitting clothing near forge.

8. Use only the right tongs for the job.

9. Turn on ventilation fan before igniting forge and do not turn off until forge has been turned off.

10. Have fire extinguisher, fire blanket, and first aid kit readily available.

D. **COMPLETION QUESTIONS:**

1. Before starting the gas forge ignition procedure, the _____ _____ should be turned on.

2. At the start of the ignition procedure, the gas control should be _____ and the air control _____ _____.

3. In shutting down the forge, the _____ control should be closed first.

4. The height of the flame should be just above the _____.

5. Do not reignite the forge within _____ minutes of shutdown.

6. The proper flame shape resembles a _____ _____.

7. Flame height is adjusted with the _____.

8. Minor _____ adjustments may be made by rotating the disc located near the blower inlet.

9. In shutting down the manually ignited forge, first turn _____ to the "off" position, close the _____ _____, and then stop the _____.

10. Failure to stop the _____ could result in cracking of the firebrick.

NOTES

ARC WELDER

A. PART IDENTIFICATION:

Identify the circled parts on the arc welder illustrated below.

1. _____
2. _____
3. _____
4. _____
5. _____
6. _____
7. _____
8. _____
9. _____
10. _____
11. _____
12. _____

B. SAFE OPERATIONAL PROCEDURES:

1. Electrode selection:

    a. The size of electrode to use is determined by the thickness of the metal. The most common sizes used are 1/8" and 5/32".

    b. The type of electrode is determined by the metal to be welded and the type of welder used.

Metal to be Welded	A.W.S. Number	Welder
Clean, mild steel	E6013	AC or DC-SP
Clean, mild steel	E6024	AC or DC-RP
Rusty, dirty steel	E6011	AC or DC-RP
Higher carbon steel	E7018	DC-RP or AC

    c. American Welding Society (A.W.S.) numbers are given to electrodes so that electrodes made by different manufacturers can be compared on the same basis.

2. Adjusting the welder and striking the arc:

   a. Select the proper electrode considering the thickness and type of metal to be welded and the type of welder.

   b. Set the amperage at about 115 amps for the 1/8" electrode and about 103 amps for the 5/32" size.

   c. Position a piece of metal on the welding table and select the proper arc welding helmet.

   d. With the ground clamp connected to the table and the helmet in position, strike the arc as a match would be struck.

   e. Raise the electrode about 1/2" from the metal as the arc is established; then slowly lower the electrode to about 1/8" from the metal as a molten puddle is formed.

   f. Repeat "d" and "e" until 10 molten puddles can be established without sticking the electrode.

   g. If the electrode sticks, first attempt to twist it free. If this fails, release the holder, turn off the switch, and use a pliers to remove the electrode.

   h. Never turn off the switch while current is flowing through the electrode holder. This would arc the switch.

3. Welding a bead:

   a. Follow the steps in striking an arc.

   b. When the arc is established, tilt the electrode 15° to 20° in the direction of travel (from left to right for right-handed welders).

   c. Move the electrode ahead at a smooth and uniform rate.

   d. Keep the electrode at the forward edge of the crater and feed it into the work as it melts off to maintain the proper arc length.

   e. The width of the bead should be 1-1/2 times the width of the electrode.

C. GENERAL SAFETY PRACTICES:

1. Always wear a welding helmet with at least a number 10 lens when operating an arc welder.

2. Do not operate the arc welder without permission from the instructor.

3. Wear long sleeves and gloves. Avoid cuffs on trousers and perforated or low-top shoes.

4. Avoid exposure of any skin surface when operating an arc welder.

5. Wear safety glasses at all times when operating an arc welder.

6. Perform work on concrete or fireproof surfaces.

7. Give the word "cover" to all people standing nearby when ready to strike an arc.

8. Always avoid looking at the arc with the unprotected eye.

9. Never weld on a container until it has been made safe.

10. All safety practices that apply to arc welding also apply to the carbon arc torch.

11. Do not weld while wearing dirty or oil-soaked clothing.

12. Do not lay the electrode holder on a grounded table.

13. Do not stand on a wet floor while welding.

14. When making repairs or adjustments on the welder, always open the main switch.

15. Clean the floor and table of scraps before starting to weld.

16. Do not weld without proper ventilation.

17. Do not weld galvanized metal indoors; a toxic gas is given off.

18. Check welding cables for damaged insulation.

19. Use caution in releasing the ground clamp after welding; it may be hot due to poor contact with the table.

20. Do not use water to extinguish fires near the welder. Use the dry chemical fire extinguisher.

21. Do not use a welder that does not have its case properly grounded.

22. Do not leave hot metal where others might come in contact with it.

23. Do not adjust the welder while it is in operation.

24. Never weld with a cracked lens in the helmet.

25. Handle hot metal with a pliers or tongs.

26. Keep welding tools in their proper location.

27. Protect others with the welding screen whenever possible.

D. **COMPLETION QUESTIONS:**

1. A.W.S. stands for _____ _____ _____.

2. The proper arc length when welding is about _____ inch(es).

3. The arc is struck like striking a _____.

4. The angle the electrode is tilted in the direction of travel is _____ ° to _____ °.

5. Two common sizes of electrodes used are _____ and _____.

6. The width of the bead should be _____ times the diameter of the electrode.

7. _____ _____ determines the size of electrode to use.

8. If the electrode sticks to the work, first attempt to _____ it free.

9. Never _____ _____ the switch while current is flowing through the electrode holder.

10. While welding a bead, keep the electrode at the _____ edge of the crater.

OXYACETYLENE WELDER

A. PART IDENTIFICATION:

Identify the circled parts on the oxyacetylene welder illustrated below.

1. _____
2. _____
3. _____
4. _____
5. _____
6. _____
7. _____
8. _____
9. _____
10. _____
11. _____
12. _____
13. _____
14. _____
15. _____
16. _____
17. _____
18. _____

B. SAFE OPERATIONAL PROCEDURES:

1. Setting up equipment:

    a. Fasten oxygen and acetylene cylinders securely in an upright position.

    b. "Crack" the cylinder valves to blow out dust.

    c. Attach the regulators to the cylinder valves. Some acetylene regulator nuts have left-hand threads.

    d. Be sure the regulator nuts fit the cylinder valves properly. Do not force the threads.

e. Be sure the regulator valves are off. Turn the adjusting screws counter-clockwise until they are loose in the threads.

f. Open the oxygen cylinder valve very slowly until the gage reaches its maximum reading; then turn the cylinder valve all the way open.

g. Open the acetylene cylinder valve slowly, only one-half turn.

h. Test connections suspected of leaking with nonoil-based soap suds.

2. Lighting the torch:

a. Open acetylene cylinder valve one-half turn and leave the wrench in position at all times.

b. Open the oxygen cylinder valve all the way open.

c. Open the regulator valves by turning the adjusting screw clockwise until the proper pressure is obtained for the job.

d. Open the acetylene valve at the torch 1/8 turn.

e. Use a friction lighter and light the torch.

f. Adjust the acetylene valve until the flame is just ready to leave the tip of the torch.

g. Place the appropriate shaded goggles or face shield over the eyes.

h. Open oxygen torch valve until a neutral flame is produced. A neutral flame has no feather and has a blunt or rounded inner cone.

3. Cutting with oxyacetylene:

a. Select the proper hose pressure for the job. Common pressures for cutting range from 30 to 40 PSI for oxygen and 3 to 5 PSI for acetylene.

b. Open the blowpipe acetylene valve about 1/8 turn; then light the torch with a friction lighter.

c. Increase the acetylene until the flame is just ready to leave the tip.

d. Open the oxygen preheat valve until a neutral flame is obtained.

e. Hold the torch with both hands with the right hand near the lever.

f. Keep the inner cones of the neutral preheat flames about 1/16" above the metal.

g. Allow a molten puddle to form on the edge; then press the oxygen lever.

h. Tilt the torch so the flame leads slightly and move along as fast as possible from right to left.

4. Turning off the torch:

    a. Turn off the acetylene valve and then the oxygen valve.

    b. When closing down for a long period of time or when finished with a job,

        (1) Close the oxygen cylinder valve.

        (2) Open the blowpipe oxygen valve to release all pressure from hose and regulator.

        (3) Turn out the pressure adjusting screw of the oxygen regulator until the threads are loose.

        (4) Close the blowpipe oxygen valve.

        (5) Close the acetylene cylinder valve.

        (6) Open the blowpipe acetylene valve to release all pressure from the hose and regulator.

        (7) Turnout the pressure adjusting screw of the acetylene regulator until the threads are loose.

        (8) Close the blowpipe acetylene valve and hang up the hoses.

C. GENERAL SAFETY PRACTICES:

1. Always wear safety glasses with a shaded face shield or shaded goggles when welding or cutting.

2. Do not operate the oxyacetylene welder without permission from the instructor.

3. Never cut or weld a container until it has been made safe.

4. Wear gloves and clean coveralls when welding or cutting.

5. Turn down pants cuffs and wear high-top shoes when cutting or welding.

6. Do not cut galvanized metal indoors. A toxic gas is given off.

7. Never work with oxyacetylene equipment that may be defective.

8. Never use oil or grease around the oxygen cylinder, regulator, or torch.

9. Inspect hoses and test the connections for leaks using only nonoil-based soap suds.

10. Never repair a hose with tape.

11. Never light the torch with a match.

12. Keep hot metal and the flame away from the hoses.

13. Keep the protector caps on cylinders not in use. Never drop cylinders.

14. Never use acetylene pressures above 15 PSI.

15. Never open acetylene cylinder valves more than 1/2 turn. This will allow the cylinder to be shut off quickly in the event of a fire.

16. Never release oxygen or acetylene in a confined area. The acetylene may ignite and the concentrated oxygen may cause clothing and other combustible material to burn very fast.

17. Never point the torch toward anyone.

18. Never use oxygen or acetylene to blow dirt off clothing. Clothing saturated with oxygen or acetylene will burn very fast.

19. If a flashback should occur, turn off the oxygen valve on the torch and close the acetylene cylinder valve immediately. Cool the torch and determine the cause of the flashback.

20. Clear the area of all combustible materials before lighting the torch.

21. Never leave hot metal where others may be burned by it.

22. Turn on the ventilation fan before lighting the oxyacetylene welder.

D. COMPLETION QUESTIONS:

1. Oxygen and acetylene cylinders should be securely fastened in a _____ position.

2. _____ the cylinder valves to blow out the dust before attaching regulators.

3. The acetylene regulator nut has _____ hand threads.

4. In turning off the regulators, the adjusting screws must be turned _____.

5. The acetylene cylinder valve should be opened only _____ turn(s).

6. Use _____ _____ to test connections for leaks.

7. A neutral flame has no _____ and has a _____ inner cone.

8. Before lighting the torch, the blowpipe acetylene valve should be opened about _____ turn.

9. The _____ valve should be shut off first to turn off the torch.

10. When cutting with the oxyacetylene torch, the _____ oxidizes the metal after it reaches the kindling temperature.

NOTES

NOTES

NOTES

NOTES

NOTES

NOTES

www.ingramcontent.com/pod-product-compliance
Lightning Source LLC
Chambersburg PA
CBHW060516300426
44112CB00017B/2687